Astrophysics of Black Holes

Astrophysics and Space Science Library

More information about this series at http://www.springer.com/series/5664

Cosimo Bambi

Editor

Astrophysics of Black Holes

From Fundamental Aspects to Latest
Developments

 Springer

Editor
Cosimo Bambi
Department of Physics
Fudan University
Shanghai
China

ISSN 0067-0057   ISSN 2214-7985   (electronic)
Astrophysics and Space Science Library
ISBN 978-3-662-57087-6   ISBN 978-3-662-52859-4   (eBook)
DOI 10.1007/978-3-662-52859-4

Preface

The *Fudan Winter School on Astrophysical Black Holes* was held at Fudan University, in Shanghai, from February 10 to 15, 2014. The school was organized with the spirit to contribute to the growth of the high-energy astrophysics community in China and to promote the interaction among people working in different places of the country. It seemed also to be a good way to make know the new gravity and high-energy physics group at the Department of Physics of Fudan University. Departments of physics in China have a strong tradition in condensed matter and applied physics, while they are weak in the other fields. As the rest of the country, they are now in a phase of quick expansion and they are trying to form research groups working even in other areas. High-energy astrophysics seems to be one of the most active new directions, and there are an increasing number of students who want to work in this field. However, with the exception of a few cases, departments of physics in China have not yet a satisfactory astrophysics program. Fudan University is probably one of the best examples: While it is one of the top universities in China, there is no undergraduate or postgraduate course in astronomy/astrophysics.

The school consisted of six series of lectures given by experts in their research field, covering basic topics in black hole astrophysics and important directions in current research. It was mainly intended for master/Ph.D. students. The event was attended by approximately 60 participants. At the end of the school, we considered the possibility of organizing a contributed volume, by inviting each lecturer to write an independent chapter. Most lecturers accepted the invitations, and the result is this book. This volume has all the features to become a very useful reference book to researchers and students working on astrophysical black holes.

Shanghai   Cosimo Bambi
January 2016

Contents

Contributors

Tomaso M. Belloni INAF-Osservatorio Astronomico di Brera, Merate, Italy

Jean-Pierre Lasota Institut d'Astrophysique de Paris, CNRS et Sorbonne Universités, UPMC Univ Paris 06, Paris, France; Nicolaus Copernicus Astronomical Center, Warsaw, Poland

Daniele Malafarina Department of Physics and Center for Field Theory and Particle Physics, Fudan University, Shanghai, China; Physics Department, SST, Nazarbayev University, Astana, Kazakhstan

M. Middleton Institute of Astronomy, Cambridge, UK

Sara E. Motta Department of Physics, Astrophysics, University of Oxford, Oxford, UK

Feng Yuan Shanghai Astronomical Observatory, Chinese Academy of Sciences, Shanghai, China

Chapter 1
Black Hole Accretion Discs

Jean-Pierre Lasota

Abstract This is an introduction to the models of accretion discs around black holes. After a presentation of the nonrelativistic equations describing the structure and evolution of geometrically thin accretion discs, we discuss their steady-state solutions and compare them to observation. Next, we describe in detail the thermal–viscous disc instability model and its application to dwarf novae for which it was designed and its X-ray irradiated disc version which explains the soft X-ray transients, i.e. outbursting black hole low-mass X-ray binaries. We then turn to the role of advection in accretion flows onto black holes illustrating its action and importance with a toy model describing both ADAFs and slim discs. We conclude with a presentation of the general-relativistic formalism describing accretion discs in the Kerr spacetime.

1.1 Introduction

The author of this chapter is old enough to remember the days when even serious astronomers doubted the existence of accretion discs and scientists snorted with contempt at the suggestion that there might be such things as black holes; the very possibility of their existence was rejected, and the idea of black holes was dismissed as a fancy of eccentric theorists. Today, some 50 years later, there is no doubt about the existence of accretion discs and black holes; both have been observed and shown to be ubiquitous in the universe. The spectacular ALMA image of the protostellar disc in HL Tau [46] is breathtaking, and we can soon expect to see the silhouette of a supermassive black hole in the centre of the Galaxy in near infrared [57] or millimetre waves [10].

J.-P. Lasota (✉)
Institut d'Astrophysique de Paris, CNRS et Sorbonne Universités,
UPMC Univ Paris 06, UMR 7095, 98bis Bd Arago, 75014 Paris, France
e-mail: lasota@iap.fr

J.-P. Lasota
Nicolaus Copernicus Astronomical Center, Bartycka 18, 00-716 Warsaw, Poland

© Springer-Verlag Berlin Heidelberg 2016   1
C. Bambi (ed.), *Astrophysics of Black Holes*, Astrophysics
and Space Science Library 440, DOI 10.1007/978-3-662-52859-4_1

Understanding accretion discs around black holes is interesting in itself because of the fascinating and complex physics involved but is also fundamental for understanding the coupled evolution of galaxies and their nuclear black holes, i.e. fundamental for the understanding the growth of structures in the universe. The chance that inflows onto black holes are strictly radial, as assumed in many models, is slim.

The aim of the present chapter was to introduce the reader to models of accretion discs around black holes. Because of the smallness of black holes, the sizes of their accretion discs span several orders of magnitude: from close to the horizon up 100 000 or even 1 000 000 black hole radii. This implies, for example, that the temperature in a disc around a stellar-mass black hole varies from 10^7 K, near the its surface, to $\sim 10^3$ K near the disc's outer edge at 10^5 black hole radii, say. Thus, studying black hole accretion discs allows the study of physical regimes relevant also in a different context and, inversely, the knowledge of accretion disc physics in other systems, such as protostellar discs or cataclysmic variable stars, is useful or even necessary for understanding the discs around black holes.

Section 1.2 contains a short discussion of the disc-driving mechanisms and introduces the α-prescription used in this chapter. In Sect. 1.3 after presenting the general framework of the geometrically thin-disc model, we discuss the properties of stationary solutions and the Shakura–Sunyaev solution in particular. The dwarf nova disc instability model and its application to black hole transient sources are the subject of Sect. 1.4. The role of advection in accretion onto black holes is presented in Sect. 1.5 with the main stress put on high accretion rate flows. Finally, Sects. 1.6 and 1.7 about the general-relativistic version of the accretion disc equations conclude the present chapter.

NOTATIONS AND DEFINITIONS
The Schwarzschild radius (radius of a non-rotating black hole) is

$$R_S = \frac{2GM}{c^2} = 2.95 \times 10^5 \frac{M}{M_\odot} \text{ cm}, \tag{1.1}$$

where M is the mass of the gravitating body and c the speed of light.
The Eddington accretion rate is defined as

$$\dot{M}_{\text{Edd}} = \frac{L_{\text{Edd}}}{\eta c^2} = \frac{1}{\eta} \frac{4\pi GM}{c\kappa_{es}} = \frac{1}{\eta} \frac{2\pi cR_S}{\kappa_{es}} = 1.6 \times 10^{18} \eta_{0.1}^{-1} \frac{M}{M_\odot} \text{ g s}^{-1}, \tag{1.2}$$

where $\eta = 0.1\eta_{0.1}$ is the radiative efficiency of accretion and κ_{es} the electron scattering (Thomson) opacity.

We will often use accretion rate measured in units of Eddington accretion rate:

$$\dot{m} = \frac{\dot{M}}{\dot{M}_{\text{Edd}}}. \tag{1.3}$$

Additional reading: There are excellent general reviews of accretion disc physics, and they can be found in references [9, 18, 26, 55].

1.2   Disc-Driving Mechanism; Viscosity

In recent years, there have been an impressive progress in understanding the physical mechanisms that drive disc accretion. It is now obvious that the turbulence in ionized Keplerian discs is due to the magnetorotational instability (MRI) also known as the Balbus–Hawley mechanisms [7, 8]. However, despite these developments, numerical simulations, even in their global 3D form, suffer still from weaknesses that make their direct application to real accretion flows almost infeasible.

One of the most serious problems is the value of the ratio of the (vertically averaged) total stress to thermal (vertically averaged) pressure

$$\alpha = \frac{\langle \tau_{r\varphi} \rangle_z}{\langle P \rangle_z} \tag{1.4}$$

which according to most MRI simulation is $\sim 10^{-3}$, whereas observations of dwarf nova decay from outburst unambiguously show that $\alpha \approx 0.1 - 0.2$ [29, 54]. Only recently Hirose et al. [22] showed that effects of convection at temperatures $\sim 10^4$ K increase α to values ~ 0.1. This might solve the problem of discrepancies between the MRI calculated and the observed value of α [11]. One has, however, to keep in mind that the simulations in question have been performed in the so-called shearing box and their validity in a generic 3D case has yet to be demonstrated.

Another problem is related to cold discs such as quiescent dwarf nova discs [31] or protostellar discs [7]. For the standard MRI to work, the degree of ionization in a weakly magnetized, quasi-Keplerian disc must be sufficiently high to produce the instability that leads to a breakdown of laminar flow into turbulence which is the source of viscosity-driving accretion onto the central body. In cold discs, the ionized fraction is very small and might be insufficient for the MRI to operate. In any case, in such a disc non-ideal MHD effects are always important. All these problems still await their solution.

Finally, and very relevant to the subject of this chapter, there is the question of stability of discs in which the pressure is due to radiation and opacity to electron scattering. According to theory, such discs should be violently (thermally) unstable, but observations of systems presumed to be in this regime totally infirm this prediction. MRI simulations not only do not solve this contradiction but rather reinforce it [24].

1.2.1 The α-Prescription

The α-prescription [53] is a rather simplistic description of the accretion disc physics, but before one is offered better and physically more reliable options, its simplicity makes it the best possible choice and has been the main source of progress in describing accretion discs in various astrophysical contexts.

One keeps in mind that the accretion-driving viscosity is of magnetic origin, but one uses an effective hydrodynamical description of the accretion flow. The hydrodynamical stress tensor is (see e.g. [34])

$$\tau_{r\varphi} = \rho v \frac{\partial v_\varphi}{\partial R} = \rho \frac{d\Omega}{d\ln R}, \qquad (1.5)$$

where ρ is the density, v the kinematic viscosity coefficient and v_φ the azimuthal velocity ($v_\varphi = R\Omega$).

In 1973, Shakura and Sunyaev proposed the (now famous) prescription

$$\tau_{r\varphi} = \alpha P, \qquad (1.6)$$

where P is the total thermal pressure and $\alpha \leq 1$. This leads to

$$v = \alpha c_s^2 \left[\frac{d\Omega}{d\ln R} \right]^{-1}, \qquad (1.7)$$

where $c_s = \sqrt{P/\rho}$ is the isothermal sound speed and ρ the density. For the Keplerian angular velocity

$$\Omega = \Omega_K = \left(\frac{GM}{R^3} \right)^{1/2} \qquad (1.8)$$

this becomes

$$v = \frac{2}{3}\alpha c_s^2/\Omega_K. \qquad (1.9)$$

Using the approximate hydrostatic equilibrium Eq. (1.19), one can write this as

$$v \approx \frac{2}{3}\alpha c_s H. \qquad (1.10)$$

Multiplying the rhs of Eq. (1.5) by the ring length ($2\pi R$) and averaging over the (total) disc height, one obtains the expression for the *total torque*

$$\mathfrak{T} = 2\pi R \Sigma v R \frac{d\Omega}{d\ln R}, \qquad (1.11)$$

where

$$\Sigma = \int_{-\infty}^{+\infty} \rho \, dz. \tag{1.12}$$

For a Keplerian disc

$$\mathfrak{T} = 3\pi \, \Sigma \nu \ell_K, \tag{1.13}$$

($\ell_K = R^2 \Omega_K$ is the Keplerian *specific* angular momentum.)

The viscous heating is proportional to $\tau_{r\varphi}(d\Omega/dR)$ [34]. In particular, the viscous heating rate per unit volume is

$$q^+ = -\tau_{r\varphi} \frac{d\Omega}{d \ln R}, \tag{1.14}$$

which for a Keplerian disc, using Eq. (1.6), can be written as

$$q^+ = \frac{3}{2} \alpha \Omega_K P, \tag{1.15}$$

and the viscous heating rate per unit surface is therefore

$$Q^+ = \frac{\mathfrak{T}\Omega'}{4\pi R} = \frac{9}{8} \Sigma \nu \Omega_K^2. \tag{1.16}$$

(The denominator in the first rhs is $2 \times 2\pi R$ taking into account the existence of two disc surfaces.)

Additional reading: References [7, 8, 22].

1.3   Geometrically Thin Keplerian Discs

The 2D structure of geometrically thin, non-self-gravitating, axially symmetric accretion discs can be split into a $1 + 1$ structure corresponding to a hydrostatic vertical configuration and radial quasi-Keplerian viscous flow. The two 1D structures are coupled through the viscosity mechanism transporting angular momentum and providing the local release of gravitational energy.

1.3.1   Disc Vertical Structure

The vertical structure can be treated as a one-dimensional star with two essential differences:

1. the energy sources are distributed over the whole height of the disc, while in a star, there limited to the nucleus,

2. the gravitational acceleration *increases* with height because it is given by the tidal gravity of the accretor, while in stars, it decreases as the inverse square of the distance from the centre.

Taking these differences into account, the standard stellar structure equations (see e.g. [48]) adapted to the description of the disc vertical structure are listed below.

- Hydrostatic equilibrium
  The gravity force is counteracted by the force produced by the pressure gradient:

$$\frac{dP}{dz} = \rho g_z, \tag{1.17}$$

where g_z is the vertical component (tidal) of the accreting body gravitational acceleration:

$$g_z = \frac{\partial}{\partial z}\left[\frac{GM}{(R^2 + z^2)^{1/2}}\right] \approx \frac{GM}{R^2}\frac{z}{R}. \tag{1.18}$$

The second equality follows from the assumption that $z \ll R$. Denoting the typical (pressure or density) scale height by H, the condition of geometrical thinness of the disc is $H/R \ll 1$, and writing $dP/dz \sim P/H$, Eq. (1.17) can be written as

$$\frac{H}{R} \approx \frac{c_s}{v_K}, \tag{1.19}$$

where $v_K = \sqrt{GM/R}$ is the Keplerian velocity, and we made use of Eq. (1.18). From Eq. (1.19), it follows that

$$\frac{H}{c_s} \approx \frac{1}{\Omega_K} =: t_{\mathrm{dyn}}, \tag{1.20}$$

where t_{dyn} is the dynamical time.
- Mass conservation
  In 1D hydrostatic equilibrium, the mass conservation equation takes the simple form of

$$\frac{d\varsigma}{dz} = 2\rho. \tag{1.21}$$

- Energy transfer—temperature gradient

$$\frac{d \ln T}{dz} = \nabla \frac{d \ln P}{dz}. \tag{1.22}$$

For radiative energy transport

$$\nabla_{\mathrm{rad}} = \frac{\kappa_R P F_z}{4P_r c g_z}, \tag{1.23}$$

where P_r is the radiation pressure and κ_R the Rosseland mean opacity. From Eqs. (1.22) and (1.23), one recovers the familiar expression for the radiative flux

$$F_z = -\frac{16}{3}\frac{\sigma T^3}{\kappa_R \rho}\frac{\partial T}{\partial z} = -\frac{4\sigma}{3\kappa_R \rho}\frac{\partial T^4}{\partial z} \tag{1.24}$$

(F_z is positive because the temperature decreases with z so $\partial T/\partial z < 0$).
The photosphere is at optical thickness $\tau \simeq 2/3$ (see Eq. 1.74). The boundary conditions are as follows: $z = 0$, $F_z = 0$, $T = T_c$, and $\varsigma = 0$ at the disc midplane; at the disc photosphere, $\varsigma = \Sigma$ and $T^4(\tau = 2/3) = T_{\text{eff}}^4$. For a detailed discussion of radiative transfer, temperature stratification and boundary conditions see Sect. 1.3.5.
In the same spirit as Eq. (1.19), one can write Eq. (1.24) as

$$F_z \approx \frac{4}{3}\frac{\sigma T_c^4}{\kappa_R \rho H} = \frac{8}{3}\frac{\sigma T_c^4}{\kappa_R \Sigma}, \tag{1.25}$$

where T_c is the midplane ("central") disc temperature. Using the optical depth $\tau = \kappa_R \rho H = (1/2)\kappa_R \Sigma$, this can be written as

$$F_z(H) \approx \frac{8}{3}\frac{\sigma T_c^4}{\tau} = Q^-, \tag{1.26}$$

(see Eq. 1.77 for a rigorous derivation of this formula).

Remark 1.1   In some references (e.g. in [18]), the numerical factor on the rhs is "4/3" instead of "8/3". This is due to a different definition of Σ: in our case, it is $= 2\rho H$, whereas in [18] $\Sigma = \rho H$.

In the case of convective energy transport, $\nabla = \nabla_{\text{conv}}$. Because convection in discs is still not well understood (see, however, [22]), there is no obvious choice for ∇_{conv}. In practice, a prescription designed by Paczyński [41] for extended stellar envelope is used [21], but this most probably does not represent very accurately what is happening in convective accretion discs [11].

- Energy conservation
  Vertical energy conservation should have the form

$$\frac{dF_z}{dz} = q^+(z), \tag{1.27}$$

where $q^+(z)$ corresponds to viscous energy dissipation per unit volume.

Remark 1.2 In contrast with accretion discs, stellar envelopes have $dF_z/dz = 0$

The α-prescription does not allow deducing the viscous dissipation stratification (z dependence), and it just says that the vertically averaged viscous torque is proportional to pressure. Most often one assumes therefore that

$$q^+(z) = \frac{3}{2}\alpha\Omega_{\mathrm{K}}P(z),\qquad(1.28)$$

by analogy with Eq. (1.15), but such an assumption is chosen because of its simplicity and not because of some physical motivation. In fact, MRI numerical simulations suggest that dissipation is not stratified in the way is pressure [22].

- The vertical structure equations have to be completed by the equation of state (EOS):

$$P = P_r + P_g = \frac{4\sigma}{3c}T^4 + \frac{\mathscr{R}}{\mu}\rho T,\qquad(1.29)$$

where $\mathscr{R}$ is the gas constant and μ the mean molecular weight, and an equation describing the mean opacity dependence on density and temperature.

1.3.2 Disc Radial Structure

- Continuity (mass conservation) equation has the form

$$\frac{\partial \Sigma}{\partial t} = -\frac{1}{R}\frac{\partial}{\partial R}(R\Sigma v_{\mathrm{r}}) + \frac{S(R,t)}{2\pi R},\qquad(1.30)$$

where $S(R, t)$ is the matter source (sink) term.
In the case of an accretion disc in a binary system,

$$S(R, t) = \frac{\partial \dot{M}_{\mathrm{ext}}(R, t)}{\partial R}\qquad(1.31)$$

represents the matter brought to the disc from the Roche lobe filling/mass losing (secondary) companion of the accreting object. $\dot{M}_{\mathrm{ext}} \approx \dot{M}_{\mathrm{tr}}$, where $\dot{M}_{\mathrm{tr}}$ is the mass-transfer rate from the companion star. Most often one assumes that the transfer stream delivers the matter exactly at the outer disc edge, but although this assumption simplifies calculations, it is contradicted by observations that suggest that the stream overflows the disc surface(s).
- Angular momentum conservation

$$\frac{\partial \Sigma\ell}{\partial t} = -\frac{1}{R}\frac{\partial}{\partial R}(R\Sigma\ell v_{\mathrm{r}}) + \frac{1}{R}\frac{\partial}{\partial R}\left(R^3\Sigma v\frac{d\Omega}{dR}\right) + \frac{S_\ell(R,t)}{2\pi R}.\qquad(1.32)$$

This conservation equation reflects the fact that angular momentum is transported through the disc by a viscous stress $\tau_{r\varphi} = R\Sigma v d\Omega / dR$. Therefore, if the disc is not considered infinite (recommended in application to real processes and systems), there must be somewhere a sink of this transported angular momentum $S_\ell(R, t)$. For binary semidetached binary systems, there is both a source (angular momentum brought in by the mass-transfer stream form the stellar companion) and a sink (tidal interaction taking angular momentum back to the orbit). The two respective terms in the angular momentum equation can be written as

$$S_j(R, t) = \frac{\ell_k}{2\pi R}\frac{\partial \dot{M}_{ext}}{\partial R} - \frac{T_{tid}(R)}{2\pi R}. \tag{1.33}$$

Assuming $\Omega = \Omega_K$, from Eqs. (1.30) and (1.32), one can obtain an diffusion equation for the surface density Σ:

$$\frac{\partial \Sigma}{\partial t} = \frac{3}{R}\frac{\partial}{\partial R}\left\{R^{1/2}\frac{\partial}{\partial R}\left[v\Sigma R^{1/2}\right]\right\}. \tag{1.34}$$

Comparing with Eq. (1.30) one sees that the radial velocity induced by the viscous torque is

$$v_r = -\frac{3}{\Sigma R^{1/2}}\frac{\partial}{\partial R}\left[v\Sigma R^{1/2}\right], \tag{1.35}$$

which is an example of the general relation

$$v_{visc} \sim \frac{v}{R}. \tag{1.36}$$

Using Eq. (1.10), one can write

$$t_{vis} := \frac{R}{v_{visc}} \approx \frac{R^2}{v} \approx \alpha^{-1}\frac{H}{c_s}\left(\frac{H}{R}\right)^{-2}. \tag{1.37}$$

The relation between the viscous and the dynamical times is

$$t_{vis} \approx \alpha^{-1}\left(\frac{H}{R}\right)^{-2} t_{dyn}. \tag{1.38}$$

In thin ($H/R \ll 1$) accretion discs, the viscous time is much longer that the dynamical time. In other words, during viscous processes, the vertical disc structure can be considered to be in hydrostatic equilibrium.

• Energy conservation
  The general form of energy conservation (thermal) equation can be written as
  follows:

$$\rho T \frac{ds}{dR} := \rho T \frac{\partial s}{\partial t} + v_r \frac{\partial s}{\partial R} = q^+ - q^- + \tilde{q}, \qquad (1.39)$$

where s is the entropy density, q^+ and q^- are the viscous and radiative energy
density, respectively, and $\tilde{q}$ is the density of external and/or radially transported
energy densities. Using the first law of thermodynamics $Tds = dU + PdV$, one
can write

$$\rho T \frac{ds}{dt} = \rho \frac{dU}{dt} + P \frac{\partial v_r}{\partial r}, \qquad (1.40)$$

where $U = \Re T_c / \mu (\gamma - 1)$.
Vertically averaging, but taking $T = T_c$, using Eq. (1.30) and the thermodynamical
relations from Appendix (for $\beta = 1$), one obtains

$$\frac{\partial T_c}{\partial t} + v_r \frac{\partial T_c}{\partial R} + \frac{\Re T_c}{\mu c_P} \frac{1}{R} \frac{\partial (R v_r)}{\partial R} = 2 \frac{Q^+ - Q^-}{c_P \Sigma} + \frac{\tilde{Q}}{c_P \Sigma}, \qquad (1.41)$$

where Q^+ and Q^- are the heating and cooling rates per unit surface, respectively.
$\tilde{Q} = Q_{\text{out}} + J$ with Q_{out} corresponding to energy contributions by the mass-transfer
stream and tidal torques; $J(T, \Sigma)$ represent radial energy fluxes that are a more
or less ad hoc addition to the 1+1 scheme to which they do not belong since it
assumes that radial gradients $(\partial/\partial R)$ of physical quantities can be neglected.
The viscous heating rate per unit surface can be written as (see Eq. 1.16)

$$Q^+ = \frac{9}{8} \nu \Sigma \Omega_K^2 \qquad (1.42)$$

while the cooling rate over unit surface (the radiative flux) is obviously

$$Q^- = \sigma T_{\text{eff}}^4. \qquad (1.43)$$

In thermal equilibrium, one has

$$Q^+ = Q^-. \qquad (1.44)$$

The cooling time can be easily estimated from Eq. (1.44). The energy density to be
radiated away is $\sim \rho c_s^2$ (see Eqs. 1.228 and 1.232), so the energy per unit surface
is $\sim \Sigma c_s^2$ and the cooling (thermal) time is

$$t_{\text{th}} = \frac{\Sigma c_s^2}{Q^-} = \frac{\Sigma c_s^2}{Q^+} \sim \alpha^{-1} \Omega_K^{-1} = \alpha^{-1} t_{\text{dyn}}. \qquad (1.45)$$

Since $\alpha < 1$, $t_{th} > t_{dyn}$ and during thermal processes, the disc can be assumed to be in (vertical) hydrostatic equilibrium.

For geometrical thin ($H/R \ll 1$) accretion discs, one has the following hierarchy of the characteristic times

$$t_{dyn} < t_{th} \ll t_{vis}. \tag{1.46}$$

(This hierarchy is similar to that of characteristic times in stars: the dynamical is shorter than the thermal (Kelvin–Helmholtz) and the thermal is much shorter than the thermonuclear time.)

1.3.3 Self-gravity

In this chapter, we are interested in discs that are not self-gravitating, i.e. in discs where the vertical hydrostatic equilibrium is maintained against the pull of the accreting body's tidal gravity, whereas the disc's self-gravity can be neglected. We will see now under what conditions this assumption is satisfied.

The equation of vertical hydrostatic equilibrium can be written as

$$\frac{1}{\rho}\frac{dP}{dz} = -g = (-g_z - g_s) = g_z\left(1 + \frac{g_s}{g_z}\right) =: -g_z\left(1 + A\right), \tag{1.47}$$

and therefore, self-gravity is negligible when $A \ll 1$. Treating the disc as an infinite uniform plane (i.e. assuming the surface density does not vary too much with radius), one can write its self-gravity as $g_s = 2\pi G\Sigma$, whereas the z-component of the gravity provided by the central body is $g_z = \Omega_K^2 z$ (Eq. 1.18). Therefore, evaluating A at $z = H$ one gets

$$A_H := \left.\frac{g_s}{g_z}\right|_H = \frac{2\pi G\Sigma}{\Omega_K^2 H}. \tag{1.48}$$

A_H is related to the so-called Toomre parameter [56]

$$Q_T := \frac{c_s\Omega}{\pi G\Sigma}, \tag{1.49}$$

widely used in the studies of gravitational stability of rotating systems, through $A_H \approx Q_T^{-1}$. We will therefore express the condition of negligible self-gravity (gravitational stability) as

$$Q_T > 1. \tag{1.50}$$

Using Eqs. (1.19), (1.10) and (1.57), one can write the Toomre parameter as

$$Q_T = \frac{3c_s^3}{G\dot{M}}, \tag{1.51}$$

or as function of the midplane temperature $T = 10^4\, T_4$

$$Q_T \approx 4.6 \times 10^7 \frac{\alpha\, T_4^{3/2}}{m\,\dot{m}}, \tag{1.52}$$

where $m = M/M_\odot$. This shows that hot ionized ($T \gtrsim 10^4$) discs become self-gravitating for high accretor masses and high accretion rates. Discs in close binary systems ($m \lesssim 30$) are never self-gravitating for realistic accretion rates ($\dot{m} < 1000$, say) and even in IMBH binaries (if they exist) (hot) discs would also be free of gravitational instability. Around a supermassive black hole, however, discs can become self-gravitating quite close to the black hole. For example, when the black hole mass is $m = 10^8$ a hot disc will become self-gravitating at $R/R_S \approx 100$, for $\dot{m} \sim 10^{-2}$. In general, geometrically thin, non-self-gravitating accretion discs around supermassive black holes have very a limited radial extent.

Additional reading: References [12, 19, 20, 35, 43, 56].

1.3.4 Stationary Discs

In the case of stationary ($\partial/\partial t = 0$) discs, Eq. (1.30) can be easily integrated giving

$$\dot{M} := 2\pi R \Sigma v_{\mathrm{r}}, \tag{1.53}$$

where the integration constant $\dot{M}$ (mass/time) is the *accretion rate*.

Also, the angular momentum equation (1.32) can be integrated to give

$$-2\pi R \Sigma v_r \ell + 2\pi R^3 \Sigma \nu \Omega' = const. \tag{1.54}$$

Or, using Eq. (1.53),

$$-\dot{M}\ell + \mathfrak{T} = const., \tag{1.55}$$

where the torque $\mathfrak{T} := 2\pi R^3 \Sigma \nu \Omega'$ (for a Keplerian disc $\mathfrak{T} = 3\pi R^2 \Sigma \nu \Omega_K$).

Assuming that at the inner disc radius, the torque vanishes one gets $const. = -\dot{M}\ell_{\mathrm{in}}$, where ℓ_{in} is the specific angular momentum at the disc inner edge. Therefore,

$$\dot{M}(\ell - \ell_{\mathrm{in}}) = \mathfrak{T} \tag{1.56}$$

which is a simple expression of angular momentum conservation.

For Keplerian discs, one obtains an important relation between viscosity and accretion rate

$$\nu \Sigma = \frac{\dot{M}}{3\pi} \left[1 - \left(\frac{R_{in}}{R} \right)^{1/2} \right]. \tag{1.57}$$

From Eqs. (1.57), (1.42) and (1.43), and the thermal equilibrium equation (1.44), it follows that

$$\sigma T_{eff}^4 = \frac{8}{3} \frac{\sigma T_c^4}{\tau} = \frac{3}{8\pi} \frac{GM\dot{M}}{R^3} \left[1 - \left(\frac{R_{in}}{R} \right)^{1/2} \right]. \tag{1.58}$$

This relation assumes only that the disc is Keplerian and in thermal ($Q^+ = Q^-$) and viscous ($\dot{M} = const.$) equilibrium. The viscosity coefficient is absent because of the thermal equilibrium assumption: in such a state, the emitted radiation flux cannot contain information about the heating mechanism, and it only says that such a mechanism exists. Steady discs do not provide information about the viscosity operating in discs or the viscosity parameter α. To get this information, one must consider (and observe) time-dependent states of accretion discs.

From Eq. (1.58) one obtains a universal temperature profile for *stationary Keplerian accretion discs*

$$T_{eff} \sim R^{-3/4}. \tag{1.59}$$

For an optically thick disc, the observed temperature $T \sim T_{eff}$ and $T \sim R^{-3/4}$ should be observed if stationary, optically thick Keplerian discs exist in the universe. And vice versa, if they are observed, this proves that such discs exist not only on paper. In 1985, Horne and Cook [23] presented the observational proof of existence of Keplerian discs when they observed the dwarf nova binary system ZCha during outburst (see Fig. 1.1).

1.3.4.1   Total Luminosity

The total luminosity of a stationary, geometrically thin accretion disc, i.e. the sum of luminosities of its two surfaces, is

$$2 \int_{R_{in}}^{R_{out}} \sigma T_{eff}^4 \, 2\pi R dR = \frac{3GM\dot{M}}{2} \int_{R_{in}}^{R_{out}} \left[1 - \left(\frac{R_{in}}{R} \right)^{1/2} \right] \frac{dR}{R^2}. \tag{1.60}$$

For $R_{out} \to \infty$, this becomes

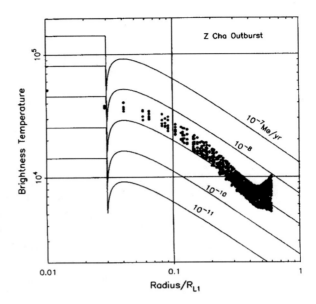

Fig. 1.1 The observed temperature profile of the accretion disc of the dwarf nova Z Cha in outburst. Near the outburst maximum such a disc is in quasi-equilibrium. The observed profile, represented by *dots* (pixels), is compared with the theoretical profiles calculated from Eq. (1.58) and represented by *continuous lines*. Pixels with $R < 0.03 R_{L1}$ correspond to the surface of the accreting white dwarf whose temperature is 40 000 K. The accretion rate in the disc is $\approx 10^{-9}\, M_\odot \mathrm{y}^{-1}$. [Fig. 6 from [23]]

$$L_{\mathrm{disc}} = \frac{1}{2}\frac{GM\dot{M}}{R_{\mathrm{in}}} = \frac{1}{2}L_{\mathrm{acc}}. \qquad (1.61)$$

In the disc, the radiating particles move on Keplerian orbits; hence, they retain half of the potential energy. If the accreting body is a black hole, this leftover energy will be lost (in this case, however, the nonrelativistic formula of Eq. 1.61 does not apply—see Eq. 1.178). In all the other cases, the leftover energy will be released in the boundary layer, if any, and at the surface of the accretor, from where it will be radiated away (Fig. 1.2).

The factor "3" in the rhs of Eq. (1.58) shows that radiation by a given ring in the accretion disc does not come only from local energy release. Indeed, in a ring between R and $R + dR$ only

$$\frac{GM\dot{M}dR}{2R^2} \qquad (1.62)$$

is being released, while

$$2 \times 2\pi R\, Q^+ dR = \frac{3GM\dot{M}}{2R^2}\left[1 - \left(\frac{R_{\mathrm{in}}}{R}\right)^{1/2}\right] dR \qquad (1.63)$$

is the total energy release. Therefore, the rest

$$\frac{GM\dot{M}}{R^2}\left[1 - \frac{3}{2}\left(\frac{R_{in}}{R}\right)^{1/2}\right] dR \qquad (1.64)$$

must diffuse out from smaller radii. This shows that viscous energy transport redistributes energy release in the disc.

1.3.5 Radiative Structure

Here, we will show an example of the solution for the vertical thin disc structure which exhibits properties impossible to identify when the structure is vertically averaged. We will also consider here an irradiated disc—such discs are present in X-ray sources.

We write the energy conservation as:

$$\frac{dF}{dz} = q^+(R, z), \qquad (1.65)$$

where F is the vertical (in the z direction) radiative flux and $q^+(R, z)$ is the viscous heating rate per unit volume. Equation (1.65) states that an accretion disc is not in radiative equilibrium ($dF/dz \neq 0$), contrary to a stellar atmosphere. For this equation to be solved, the function $q^+(R, z)$ must be known. As explained and discussed in Sect. 1.3.1, the viscous dissipation is often written as

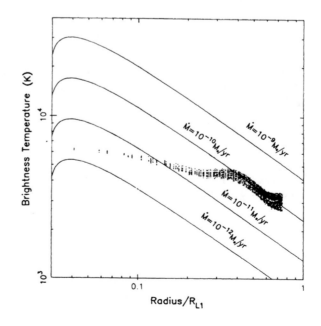

Fig. 1.2 The observed temperature profile of the accretion disc of the dwarf nova Z Cha in quiescence. This one of the most misunderstood figures in astrophysics (see text). In quiescence, the disc is *not* in equilibrium. The flat temperature profile is *exactly* what the disc instability model predicts: in quiescence, the disc temperature *must* be everywhere lower than the critical temperature, but this temperature is almost independent of the radius (see Eq. 1.89) [Fig. 11 from [58]]

$$q^+(R, z) = \frac{3}{2}\alpha\Omega_K P(z). \tag{1.66}$$

Viscous heating of this form has important implications for the structure of optically thin layers of accretion discs and may lead to the creation of coronae and winds. In reality, it is an *ad hoc* formula inspired by Eq. (1.15). We do not know yet (see, however, [11]) how to describe the viscous heating stratification in an accretion disc and Eq. (1.66) just *assumes* that it is proportional to pressure. It is simple and convenient, but it is not necessarily true.

When integrated over z, the rhs of Eq. (1.65) using Eq. (1.66) is equal to viscous dissipation per unit surface:

$$F^+ = \frac{3}{2}\alpha\Omega_K \int_0^{+\infty} P dz, \tag{1.67}$$

where $F^+ = (1/2)Q^+$ because of the integration from 0 to $+\infty$, while Q^+ contains Σ which is integrated from $-\infty$ to $+\infty$ (Eq. 1.12).

One can rewrite Eq. (1.65) as

$$\frac{dF}{d\tau} = -f(\tau)\frac{F_{\text{vis}}}{\tau_{\text{tot}}}, \tag{1.68}$$

where we introduced a new variable, the optical depth $d\tau = -\kappa_R \rho dz$, κ_R being the Rosseland mean opacity and $\tau_{\text{tot}} = \int_0^{+\infty} \kappa_R \rho dz$ is the total optical depth. $f(\tau)$ is given by:

$$f(\tau) = \frac{P}{\left(\int_0^{+\infty} P dz\right)} \frac{\left(\int_0^{+\infty} \kappa_R \rho dz\right)}{\kappa_R \rho}. \tag{1.69}$$

As ρ decreases approximately exponentially, $f(\tau)$ is the ratio of two rather well defined scale heights, the pressure and the opacity scale heights, which are comparable, so that f is of order of unity.

At the disc midplane, by symmetry, the flux must vanish: $F(\tau_{\text{tot}}) = 0$, whereas at the surface, $(\tau = 0)$

$$F(0) \equiv \sigma T_{\text{eff}}^4 = F^+. \tag{1.70}$$

Equation (1.70) states that the total flux at the surface is equal to the energy dissipated by viscosity (per unit time and unit surface). The solution of Eq. (1.68) is thus

$$F(\tau) = F^+ \left(1 - \frac{\int_0^\tau f(\tau)d\tau}{\tau_{\text{tot}}}\right), \tag{1.71}$$

where $\int_0^{\tau_{\text{tot}}} f(\tau)d\tau = \tau_{\text{tot}}$. Given that f is of order of unity, putting $f(\tau) = 1$ is a reasonable approximation. The precise form of $f(\tau)$ is more complex and is given by the functional dependence of the opacities on density and temperature; it is of no importance in this example. We thus take:

$$F(\tau) = F^+ \left(1 - \frac{\tau}{\tau_{\text{tot}}} \right). \tag{1.72}$$

To obtain the temperature stratification, one has to solve the transfer equation. Here, we use the diffusion approximation

$$F(\tau) = \frac{4}{3} \frac{\sigma \, dT^4}{d\tau}, \tag{1.73}$$

appropriate for the optically thick discs we are dealing with. The integration of Eq. (1.73) is straightforward and gives:

$$T^4(\tau) - T^4(0) = \frac{3}{4} \tau \left(1 - \frac{\tau}{2\tau_{\text{tot}}} \right) T_{\text{eff}}^4. \tag{1.74}$$

The upper (surface) boundary condition is as follows:

$$T^4(0) = \frac{1}{2} T_{\text{eff}}^4 + T_{\text{irr}}^4, \tag{1.75}$$

where T_{irr}^4 is the irradiation temperature, which depends on r, the albedo, the height at which the energy is deposited and on the shape of the disc. In Eq. (1.75), $T(0)$ corresponds to the *emergent* flux and, as mentioned above, T_{eff} corresponds to the *total* flux ($\sigma T_{\text{eff}}^4 = Q^+$) which explains the factor 1/2 in Eq. (1.75). The temperature stratification is thus:

$$T^4(\tau) = \frac{3}{4} T_{\text{eff}}^4 \left[\tau \left(1 - \frac{\tau}{2\tau_{\text{tot}}} \right) + \frac{2}{3} \right] + T_{\text{irr}}^4. \tag{1.76}$$

For $\tau_{\text{tot}} \gg 1$, the first term on the rhs has the form familiar from the stellar atmosphere models in the Eddington approximation.

In this case at $\tau = 2/3$, one has $T(2/3) = T_{\text{eff}}$.

Also for $\tau_{\text{tot}} \gg 1$, the temperature at the disc midplane is

$$T_c^4 \equiv T^4(\tau_{\text{tot}}) = \frac{3}{8} \tau_{\text{tot}} T_{\text{eff}}^4 + T_{\text{irr}}^4. \tag{1.77}$$

It is clear, therefore, that for the disc inner structure to be dominated by irradiation and the disc to be isothermal, one must have

$$\frac{F_{\text{irr}}}{\tau_{\text{tot}}} \equiv \frac{\sigma T_{\text{irr}}^4}{\tau_{\text{tot}}} \gg F^+ \tag{1.78}$$

and not just $F_{\text{irr}} \gg F^+$ as is sometimes assumed. The difference between the two criteria is important in LMXBs since, for parameters of interest, $\tau_{\text{tot}} \gtrsim 10^2 - 10^3$ in the outer disc regions.

1.3.6 Shakura–Sunyaev Solution

In their seminal and famous paper, Shakura and Sunyaev [53], found power law stationary solutions of the simplified version of the thin-disc equations presented in Sects. 1.3.1, 1.3.2 and 1.3.4. The 8 equations for the 8 unknowns T_c, ρ, P, Σ, H, ν, τ and c_s can be written as

$$\Sigma = 2H\rho \tag{i}$$

$$H = \frac{c_s R^{3/2}}{(GM)^{1/2}} \tag{ii}$$

$$c_s = \sqrt{\frac{P}{\rho}} \tag{iii}$$

$$P = \frac{\mathscr{R}\rho T}{\mu} + \frac{4\sigma}{3c}T^4 \tag{iv}$$

$$\tau(\rho, \Sigma, T_c) = \kappa_R(\rho, T_c)\Sigma \tag{v}$$

$$\nu(\rho, \Sigma, T_c, \alpha) = \frac{2}{3}\alpha c_s H \tag{vi}$$

$$\nu\Sigma = \frac{\dot{M}}{3\pi}\left[1 - \left(\frac{R_0}{R}\right)^{1/2}\right] \tag{vii}$$

$$\frac{8}{3}\frac{\sigma T_c^4}{\tau} = \frac{3}{8\pi}\frac{GM\dot{M}}{R^3}\left[1 - \left(\frac{R_0}{R}\right)^{1/2}\right]. \tag{viii}$$

Equations (i) and (ii) correspond to vertical structure Eqs. (1.21) and (1.19), Eq. (vii) is the radial Eq. (1.57), while Eq. (viii) connects vertical to radial equations. Equation (iii) defines the sound speed, Eq. (iv) is the equation of state, and Eq. (vi) contains the information about opacities. The viscosity α parametrization introduced in [53] provides the closure of the 8 disc equations. Therefore, they can be solved for a given set of α, M, R and $\dot{M}$.

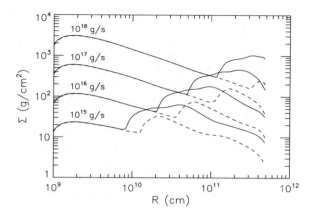

Fig. 1.3 Stationary accretion disc surface density profiles for 4 values of accretion rate. From top to bottom: $\dot{M} = 10^{18}, 10^{17}, 10^{16}$ and $10^{15} \mathrm{g\,s}^{-1}$. $m = 10\, M_{\odot}$, $\alpha = 0.1$. The *continuous line* corresponds to the unirradiated disc, the *dotted lines* to an irradiated configuration. The inner, decreasing segments of the continuous lines correspond to Eq. (1.80). *Dashed lines* describe irradiated disc equilibria (see Sect. 1.4.3) [Fig. 9 from [15]]

Power law solutions of these equations exist in physical regimes where the opacity can be represented in the Kramers form $\kappa = \kappa_0 \rho^n T^m$ and one of the two pressures, gas or radiation, dominates over the other. In [53], three regimes have been considered:

$$(a)\ P_r \gg P_g\ \text{and}\ \kappa_{\mathrm{es}} \gg \kappa_{\mathrm{ff}}$$
$$(b)\ P_g \gg P_r\ \text{and}\ \kappa_{\mathrm{es}} \gg \kappa_{\mathrm{ff}}$$
$$(c)\ P_g \gg P_r\ \text{and}\ \kappa_{\mathrm{ff}} \gg \kappa_{\mathrm{es}}.$$

Regimes (a) and (b) in which opacity is dominated by electron scattering will be discussed in Sect. 1.5. Here, we will present the solutions of regime (c), i.e. we will assume that

$$P_r = 0 \quad \text{and} \quad \kappa_R = \kappa_{\mathrm{ff}} = 5 \times 10^{24} \rho T_c^{-7/2}\ \mathrm{cm}^2\mathrm{g}^{-1}. \tag{1.79}$$

The solution for the surface density Σ, central temperature T_c and the disc relative height (aspect ratio) are, respectively,

$$\Sigma = 23\, \alpha^{-4/5} m^{1/4} R_{10}^{-3/4} \dot{M}_{17}^{7/10} f^{7/10}\ \mathrm{g\,cm}^{-2}, \tag{1.80}$$

$$T_c = 5.8 \times 10^4\, \alpha^{-1/5} m^{1/4} R_{10}^{-3/4} \dot{M}_{17}^{3/10} f^{3/10}\ \mathrm{K}, \tag{1.81}$$

$$\frac{H}{R} = 2.4 \times 10^{-2} \alpha^{-1/10} m^{-3/8} R_{10}^{1/8} \dot{M}_{17}^{3/20} f^{3/20}, \tag{1.82}$$

where $m = M/M_{\odot}$, $R_{10} = R/(10^{10}\ \mathrm{cm})$, $\dot{M}_{17} = \dot{M}/(10^{17}\ \mathrm{g\,s}^{-1})$, and $f = 1 - (R_{\mathrm{in}}/R)^{1/2}$.

Although for a 10 $M_\odot$ black hole, say, Shakura–Sunyaev solutions (1.80) and (1.81) describe discs rather far from its surface ($R \gtrsim 10^4 R_S$) the regime of physical parameters it addresses; particularly, temperatures around 10^4K are of great importance for the disc physics because it is where accretion discs become thermally and viscously unstable. This instability triggers *dwarf nova* outbursts when the accreting compact object is a white dwarf and (soft) *X-ray transients* in the case of accreting neutron stars and black holes.

It is characteristic of the Shakura–Sunyaev solution in this regime that the three Σ, T_c, and T_{eff} radial profiles vary as $R^{-3/4}$. (This implies that the optical depth τ is constant with radius—see Eq. VIII.) For high accretion rates and small radii, the assumption of opacity dominated by free–free and bound-free absorption will break down and the solution will cease to be valid. We will come to that later. Now, we will consider the other disc end: large radii.

One sees in Fig. 1.3 that for given stationary solution ($\dot{M} = const.$), the $R^{-3/4}$ slope of the Σ profiles extends down only to a minimum value $\Sigma_{\min}(R)$ after which the surface density starts to increase. With the temperature dropping below 10^4 K, the disc plasma recombines and there is a drastic change in opacities leading to a thermal instability.

Additional reading: We have assumed that accretion discs are flat. This might not be true in general because accretion discs might be warped. This has important and sometimes unexpected consequences, see, e.g. [28, 40, 45], and references therein.

1.4   Disc Instabilities

In this section, we will present and discuss the disc thermal and the (related) viscous instabilities. First, we will discuss in some detail the cause of the thermal instability due to recombination.

1.4.1   The Thermal Instability

A disc is thermally stable if radiative cooling varies faster with temperature than viscous heating. In other words,

$$\frac{d \ln \sigma T_{\mathrm{eff}}^4}{d \ln T_c} > \frac{d \ln Q^+}{d \ln T_c}. \tag{1.83}$$

Using Eq. (1.77), one obtains

$$\frac{d \ln T_{\mathrm{eff}}^4}{d \ln T_c} = 4 \left[1 - \left(\frac{T_{\mathrm{irr}}}{T_c} \right)^4 \right]^{-1} - \frac{d \ln \kappa}{d \ln T_c}. \tag{1.84}$$

In a gas pressure-dominated disc, $Q^+ \sim \rho TH \sim \Sigma T \sim T_c$. The thermal instability is due to a rapid change of opacities with temperature when hydrogen begins to recombine. At high temperatures, $d \ln \kappa / d \ln T_c \approx -4$ (see Eq. 1.79). In the instability region, the temperature exponent becomes large and positive $d \ln \kappa / d \ln T_c \approx 7 - 10$ and in the end cooling is decreasing with temperature. One can also see that irradiation by furnishing additional heat to the disc can stabilize an otherwise unstable equilibrium solution (dashed lines in Fig. 1.3).

This thermal instability is at the origin of outbursts observed in discs around black holes, neutron stars and white dwarfs. Systems containing the first two classes of objects are known as soft X-ray transients (SXTs, where "soft" relates to their X-ray spectrum), while those containing white dwarfs are called dwarf novae (despite the name that could suggest otherwise, nova and supernova outbursts have nothing to do with accretion disc outbursts).

1.4.2   Thermal Equilibria: The S-Curve

We will first consider thermal equilibria of an accretion disc in which heating is due only to local turbulence, leaving the discussion of the effects of irradiation to Sect. 1.4.3. We put therefore $T_{irr} = \widetilde{Q} = 0$. Such an assumption corresponds to discs in cataclysmic variables which are the best test bed for standard accretion disc models. The thermal equilibrium in the disc is defined by the equation $Q^- = Q^+$ (see Eq. 1.41), i.e. by

$$\sigma T_{eff}^4 = \frac{9}{8} \nu \Sigma \Omega_K^2 \tag{1.85}$$

(Equation 1.16). In general, ν is a function of density and temperature, and in the following, we will use the standard α-prescription Eq. (1.9). The energy transfer equation provides a relation between the effective and the disc midplane temperatures so that thermal equilibria can be represented as a $T_{eff}(\Sigma)$—relation (or equivalently a $\dot{M}(\Sigma)$-relation). In the temperature range of interest ($10^3 \lesssim T_{eff} \lesssim 10^5$), this relation forms an S on the (Σ, T_{eff}) plane as in Fig. 1.4. The upper, hot branch corresponds to the Shakura–Sunyaev solution presented in Sect. 1.3.6. The two other branches correspond to solutions for cold discs—along the middle branch convection plays a crucial role in the energy transfer.

Each point on the (Σ, T_{eff}) S-curve represents an accretion disc's thermal equilibrium at a given radius, i.e. a thermal equilibrium of a ring at radius R. In other words, each point of the S-curve is a solution of the $Q^+ = Q^-$ equation. Points not on the S-curve correspond to solutions of Eq. (1.41) *out of thermal equilibrium*: on the left of the equilibrium curve, cooling dominates over heating, $Q^+ < Q^-$; on the right heating over, cooling $Q^+ > Q^-$. It is easy to see that a positive slope of the $T_{eff}(\Sigma)$ curve corresponds to *stable solutions*. Indeed, a small increase of temperature of an equilibrium state (an upward perturbation) on the upper branch, say, will bring the ring to a state where $Q^+ < Q^-$ so it will cool down getting back to equilibrium. In

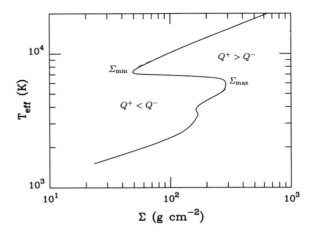

Fig. 1.4 Thermal equilibria of a ring in an accretion discs around a $m = 1.2$ white dwarf. The distance from the centre is 10^9 cm; accretion rate 6.66×10^{16} g/s. The *solid line* corresponds to $Q^+ = Q^-$. $\Sigma_{\min}$ is the critical (minimum) surface density for a hot stable equilibrium; $\Sigma_{\max}$ is the maximum surface density of a stable cold equilibrium

a similar way, an downward perturbation will provoke increased heating bringing back the system to equilibrium.

The opposite is happening along the S-curve's segment with negative slope as both temperature increase and decrease lead to a runaway. The middle branch of the S-curve corresponds therefore to *thermally unstable* equilibria.

A stable disc equilibrium can be represented only by a point on the lower, cold or the upper, hot branch of the S-curve. This means that the surface density in a stable cold state must be *lower* than the maximal value on the cold branch, $\Sigma_{\max}$, whereas the surface density in the hot stable state must be *larger* than the minimum value on this branch, $\Sigma_{\min}$. Both these critical densities are functions of the viscosity parameter α, the mass of the accreting object and the distance from the centre and depend on the disc's chemical composition. In the case of solar composition, the critical surface densities are

$$\Sigma_{\min}(R) = 39.9\, \alpha_{0.1}^{-0.80}\, R_{10}^{1.11}\, m^{-0.37} \text{ g cm}^{-2} \tag{1.86}$$

$$\Sigma_{\max}(R) = 74.6\, \alpha_{0.1}^{-0.83}\, R_{10}^{1.18}\, m_1^{-0.40} \text{ g cm}^{-2}, \tag{1.87}$$

($\alpha = 0.1\alpha_{0.1}$) and the corresponding effective temperatures are (T^+ designates the temperature at $\Sigma_{\min}$, T^- at $\Sigma_{\max}$)

$$T_{\text{eff}}^+ = 6890\, R_{10}^{-0.09}\, M_1^{0.03} \text{ K} \tag{1.88}$$

$$T_{\text{eff}}^- = 5210\, R_{10}^{-0.10}\, M_1^{0.04} \text{ K}. \tag{1.89}$$

The critical effective temperatures are practically independent of the mass and radius because they characterize the microscopic state of disc's matter (e.g. its ionization). On the other hand, the critical accretion rates depend very strongly on radius:

$$\dot{M}_{\text{crit}}^{+}(R) = 8.07 \times 10^{15} \, \alpha_{0.1}^{-0.01} \, R_{10}^{2.64} \, M_1^{-0.89} \, \text{g s}^{-1} \qquad (1.90)$$

$$\dot{M}_{\text{crit}}^{-}(R) = 2.64 \times 10^{15} \, \alpha_{0.1}^{0.01} \, R_{10}^{2.58} \, M_1^{-0.85} \, \text{g s}^{-1}. \qquad (1.91)$$

A stationary accretion disc in which there is a ring with effective temperature contained between the critical values of Eqs. (1.89) and (1.88) cannot be stable. Since the effective temperature and the surface density both decrease with radius, the stability of a disc depends on the accretion rate and the disc size (see Fig. 1.3). For a given accretion rate, a stable disc cannot have an outer radius larger than the value corresponding to Eq. (1.86).

A disc is stable if the rate (mass-transfer rate in a binary system) at which mass is brought to its outer edge ($R \sim R_d$) is larger than the critical accretion rate at this radius $\dot{M}_{\text{crit}}^{+}(R_d)$.

In general, the accretion rate and the disc size are determined by mechanisms and conditions that are exterior to the accretion process itself. In binary systems, for instance, the size of the disc is determined by the masses of the system's components and its orbital period, while the accretion rate in the disc is fixed by the rate at which the stellar companion of the accreting object loses mass, which in turn depends on the binary parameters and the evolutionary state of this stellar mass donor. Therefore, the knowledge of the orbital period and the mass-transfer rate should suffice to determine whether the accretion disc in a given interacting binary system is stable. Such knowledge allows testing the validity of the model as we will show in the next section.

1.4.2.1   Dwarf Nova and X-Ray Transient Outbursts

• Local view: the limit cycle

Let us first describe what is happening during outbursts with a disc's ring. Its states are represented by a point moving in the $\Sigma - T_{\text{eff}}$ plane as shown in Fig. 1.4 which represents accretion disc states at $R = 10^9$ cm (the accreting body has a mass of $1.2 \, M_\odot$). To follow the states of a ring during the outburst, let us start with an unstable equilibrium state on the middle, unstable branch and let us perturb it by increasing its temperature, i.e. let us shift it upwards in the $T_{\text{eff}}(\Sigma)$ plane. As we have already learned, points out of the S-curve correspond to solutions out of thermal equilibrium and in the region to the right of the S-curve heating dominates over cooling. The resulting runaway temperature increase is represented by the point moving up and reaching (in a thermal time) a quasi-equilibrium state on the hot and stable branch. It is only a *quasi*-equilibrium because the equilibrium state has been assumed to lie on the middle branch which corresponds to a lower temperature (and lower accretion rate—see Eq. 1.58). Trying to get to its proper equilibrium, the ring will cool down and move towards lower temperatures and surface densities along the upper equilibrium

Fig. 1.5 Local limit cycle of the state of disc ring at 10^9 cm during a dwarf nova outbursts. The *arrows* show the direction of motion of the system in the $T_{\text{eff}}(\Sigma)$ plane. The figure represents results of the disc instability model numerical simulations. As required by the comparison of the model with observations, the values of the viscosity parameter α on the hot and cold branches are different. [Figure adapted from [36]]

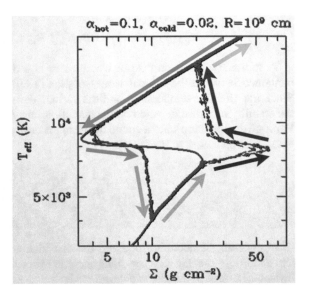

branch (in a viscous time). But the hot branch ends at $\Sigma_{\min}$, i.e. at a temperature higher (and surface density lower) than required, so the ring will never reach its equilibrium state, which is not surprising since this state is unstable. Once more the ring will find itself out of thermal equilibrium, but this time in the region where cooling dominates over heating. Rapid (thermal time) cooling will bring it to the lower cool branch. There, the temperature is lower than required so the point representing the ring will move up towards $\Sigma_{\max}$ where it will have to interrupt its (viscous time) journey having reached the end of equilibrium states before getting to the right temperature. It will find itself out of equilibrium where heating dominated over cooling so it will move back to the upper branch.

Locally, the state of a ring performing a *limit cycle* on the $\Sigma - T_{\text{eff}}$ plane moves in viscous time on the stable S-curve branches and in a thermal time between them when the ring is out of thermal equilibrium. The states on the hot branch correspond to outburst maximum and the subsequent decay, whereas the quiescent corresponds to moving on the cold branch. Since the viscosity is much larger on the hot than on the cold branch, the quiescent is much longer than the outburst phase. The full outburst behaviour can be understood only by following the whole disc evolution (Fig. 1.5).

1.4.3 Irradiation and Black Hole X-Ray Transients

We will present the global view of thermal–viscous disc outbursts for the case of X-ray transients. The main difference between accretion discs in dwarf novae and in these systems is the X-ray irradiation of the outer disc in the latter. Assuming that the irradiating X-rays are emitted by a point source at the centre of the system, one can write the irradiating flux as

$$\sigma T_{\text{irr}}^4 = \mathscr{C}\frac{L_X}{4\pi R^2} \quad \text{with} \quad L_X = \eta \min\left(\dot{M}_{\text{in}}, \dot{M}_{\text{Edd}}\right) c^2, \qquad (1.92)$$

where $\mathscr{C} = 10^{-3}\mathscr{C}_3$, η is the radiative efficiency (which can be $\ll 0.1$ for ADAFs— see below) and $\dot{M}_{\text{in}}$ is the accretion rate at the inner disc's edge. Since the physics and geometry of X-ray self-irradiation in accreting black–black hole systems are still unknown, the best we can do is to parametrize our ignorance by q constant $\mathscr{C}$ that observations suggest is $\sim 10^{-3}$. Of course, one should keep in mind that in reality, $\mathscr{C}$ might not be a constant [17].

Because the viscous heating is $\sim \dot{M}/R^3$, there always exists a radius R_{irr} for which $\sigma T_{\text{irr}}^4 > Q^+ = \sigma T_{\text{eff}}^4$. If $R_{\text{irr}} < R_d$, where R_d is the outer disc radius, the outer disc emission will be dominated by reprocessed X-ray irradiation and the structure modified as shown in Sect. 1.3.5. Irradiation will also stabilize outer disc regions (Eq. 1.84 and Fig. 1.6) allowing larger discs for a given accretion rate (see Fig. 1.3).

Irradiation modifies the critical values of the hot disc parameters:

$$\Sigma_{\text{irr}}^+ = 72.4\,\mathscr{C}_{-3}^{-0.28}\,\alpha_{0.1}^{-0.78}\,R_{11}^{0.92}\,M_1^{-0.19}\ \text{g cm}^{-2} \qquad (1.93)$$

$$T_{\text{eff}}^{\text{irr},+} = 2860\,\mathscr{C}_{-3}^{-0.09}\,\alpha_{0.1}^{0.01}\,R_{11}^{-0.15}\,M_1^{0.09}\ \text{K} \qquad (1.94)$$

$$\dot{M}_{\text{irr}}^+ = 2.3 \times 10^{17}\,\mathscr{C}_{-3}^{-0.36}\,\alpha_{0.1}^{0.04}\,R_{11}^{2.39}\,M_1^{-0.64}\ \text{g s}^{-1}. \qquad (1.95)$$

As we will see in a moment, irradiation also strongly influences the shape of outburst's light curve.

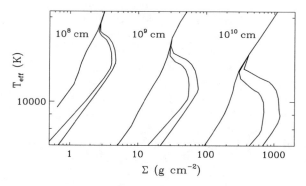

Fig. 1.6 Example S-curves for a pure helium disc with varying irradiation temperature T_{irr}. The various sets of S-curves correspond to radii $R = 10^6$, 10^9 and 10^{10} cm. For each radius, the irradiation temperature T_{irr} is 0 K, 10 000 K and 20 000 K. $\alpha = 0.16$. The instable branch disappears for high irradiation temperatures. [From [32]. Reproduced with permission from Astronomy and Astrophysics, © ESO]

* Rise to Outburst Maximum

During quiescence, the disc's surface density, temperature and accretion rate are everywhere (at all radii) on the cold branch, below their respective critical values $\Sigma_{\max}(R)$, T_{eff}^- and $\dot{M}_{\text{crit}}^-(R)$. It is important to realize that in quiescence, the disc is *not* steady: $\dot{M} \neq const$. Matter transferred from the stellar companion accumulates in the disc and is redistributed by viscosity. The surface density and temperature increase (locally, this means that the solution moves up along the lower branch of the S-curve) finally reaching their critical values. In Fig. 1.7, this happens at $\sim 10^{10}$ cm. The disc parameters entering the unstable regime triggers an outburst. In the local picture, this corresponds to leaving the lower branch of the S-curve. The next "moment" in a thermal time is represented in the left panels of Fig. 1.7. This is when a large contrast forms in the midplane temperature profile and when a surface-density spike is already above the critical line. The disc is undergoing a thermal runaway at $r \approx 8 \times 10^9$ cm. The midplane temperature rises to ~ 70000 K. This raises the viscosity which leads to an increase of the surface density, and a heating fronts start propagating inwards and outwards in the disc as shown in Fig. 1.7. In this model, the disc is truncated at an inner radius $R_{\text{in}} \approx 6 \times 10^9$ cm so the inwards propagating front quickly reaches the inner disc radius with no observable effects. It is the outwards propagating heating front that produces the outburst by heating up the disc and redistributing the mass and increasing the surface density behind it because it is also a compression front.

One should stress here that two ad hoc elements must be added to the model for it to reproduce observed outbursts of dwarf novae and X-ray transients.

- VISCOSITY. First, if the increase in viscosity were due only to the rise in the temperature through the speed of sound ($\nu \propto c_s^2$, see Eq. 1.9), the resulting outbursts would have nothing to do with the observed ones. To reproduce observed outbursts, one increases the value of α when a given ring of the disc gets to the hot branch. Ratios of hot-to-cold α of the order of 4 are used to describe dwarf nova outburst. Although in the outburst model, the α increase is an ad hoc assumption, recent MRI simulations with physical parameters corresponding to dwarf nova discs show an α increase induced by the appearance of convection [22].
- INNER TRUNCATION. Second, as mentioned already, the inner disc is assumed to be truncated in quiescence and during the rise to outburst. Although such truncation is implied and/or required by observations, its physical origin is still uncertain. The inner part of the accretion flow is of course not empty but supposed to form a $\dot{M} = const$. ADAF (see Sect. 1.5).

In our case (Fig. 1.7), the heating front reaches the outer disc radius. This corresponds to the largest outbursts. Smaller-amplitude outbursts are produced when the

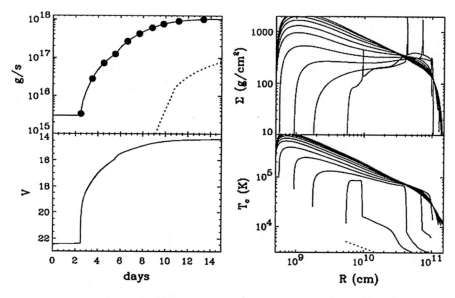

Fig. 1.7 The rise to outburst described in Sect. 1.4.3. The *upper left panel* shows $\dot{M}_{\rm in}$ and $\dot{M}_{\rm irr}$ (*dotted line*); the *bottom left panel* shows the V magnitude. Each *dot* corresponds to one of the Σ and $T_{\rm c}$ profiles in the *right panels*. The heating front propagates outwards. The disc expands during the outburst due to the angular momentum transport of the material being accreted. At $t \approx 5.5$ days, the thin disc reaches the minimum inner disc radius of the model. The profiles close to the peak are those of a steady-state disc ($\Sigma \propto T_{\rm c} \propto R^{-3/4}$). [From [16]. Reproduced with permission from Astronomy and Astrophysics, © ESO]

front does not reach the outer disc regions. In an inside-out outburst[1] the surface-density spike has to propagate uphill, against the surface-density gradient because just before the outburst $\Sigma \sim R^{1.18}$—roughly parallel to the critical surface density. Most of the mass is therefore contained in the outer disc regions. A heating front will be able to propagate if the post-front surface density is larger than $\Sigma_{\rm min}$—in other words, if it can bring successive rings of matter to the upper branch of the S-curve. If not, a *cooling front* will appear just behind the Σ spike, the heating front will die out, and the cooling front will start to propagate inwards (the heating-front will be "reflected").

The difficulty in inside-out fronts encounters when propagating is due to angular momentum conservation. In order to move outwards, the Σ spike has to take with it some angular momentum because the disc's angular momentum increases with radius. For this reason, inside-out front propagation induces a strong *outflow*. In order for matter to be accreted, a lot of it must be sent outwards. That is why during an inside-out dwarf nova outburst only $\sim 10\%$ of the disc's mass is accreted onto

---

[1] X-ray transient outbursts are always of inside-out type. In dwarf novae, both inside-out and outside-in outbursts are observed and result from calculations [31].

the white dwarf. In X-ray transients, irradiation facilitates heating front propagation (and disc emptying during decay—see next section).

The arrival of the heating front at the outer disc rim does not end the rise to maximum. After the whole disc is brought to the hot state, a surface density (and accretion rate) "excess" forms in the outer disc. The accretion rate in the inner disc corresponds to the critical one but is much higher near the outer edge. While irradiation keeps the disc hot, the excess diffuses inwards until the accretion rate is roughly constant. During this last phase of the rise to outburst maximum, $\dot{M}_{in}$ increases by a factor of 3:

$$\dot{M}_{max} \approx 3\dot{M}_{irr}^+ \approx 7.0 \times 10^{17} \mathscr{C}_{-3}^{-0.36} R_{d,11}^{2.39} m^{-0.64} \text{ g s}^{-1}. \tag{1.96}$$

Irradiation has little influence on the actual vertical structure in this region and $T_c \propto \Sigma \propto R^{-3/4}$, as in a non-irradiated steady disc. Only in the outermost disc regions does the vertical structure becomes irradiation-dominated, i.e. isothermal.

* Decay

Figure 1.8 shows the sequel to what was described in Fig. 1.7. In general, the decay from the outburst peak of an irradiated disc can be divided into three parts:

- First, X-ray irradiation of the outer disc inhibits cooling-front propagation. But since the peak accretion rate is much higher than the mass-transfer rate,[2] the disc is drained by viscous accretion of matter.
- Second, the accretion rate becomes too low for the X-ray irradiation to prevent the cooling front from propagating. The propagation speed of this front, however, is controlled by irradiation.
- Third, irradiation plays no role and the cooling front switches off the outburst on a local thermal timescale.

"Exponential Decay"

In Fig. 1.8 the "exponential decay" of the phase lasts until roughly day 80–100. At the outburst peak, the accretion rate is almost exactly constant with radius; the disc is quasi-stationary. The subsequent evolution is self-similar: the disc's radial structure evolves through a sequence of quasi-stationary ($\dot{M}(r) = const$) states. Therefore, $\nu\Sigma \sim \dot{M}_{in}(t)/3\pi$ and the total mass of the disc is thus

$$M_d = \int 2\pi R\Sigma dR \propto \dot{M}_{in} \int \frac{2}{3}\frac{r}{\nu}dr. \tag{1.97}$$

At the outburst peak, the whole disc is wholly ionized, and except for the outermost regions, its structure is very well represented by a Shakura–Sunyaev solution. In such discs, as well as in irradiation dominated discs, the viscosity coefficient satisfies the

---

[2]The peak luminosity is $\sim 3\dot{M}_{irr}^+(R_d)$, and for the disc to be unstable, the mass-transfer rate must be lower than the critical rate: $\dot{M}_{tr} < \dot{M}_{irr}^+(R_d)$.

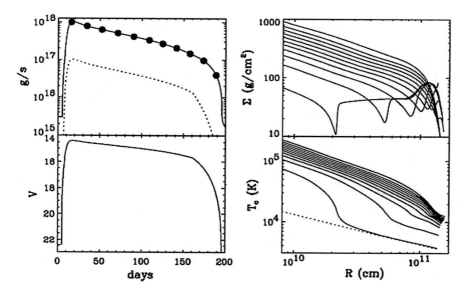

Fig. 1.8 Decay from outburst peak. The decay is controlled by irradiation until evaporation sets in at $t \approx 170$ days ($\dot{M}_{in} = \dot{M}_{evap}(R_{min})$). This cuts off irradiation, and the disc cools quickly. The irradiation cut-off happens before the cooling front can propagate through most of the disc; hence, the irradiation-controlled linear decay ($t \approx 80 - 170$ days) is not very visible in the light curve. T_{irr} (*dotted line*) is shown for the last temperature profile. [From [16]. Reproduced with permission from Astronomy and Astrophysics, © ESO]

relation $\nu \propto T \propto \dot{M}^{\beta/(1+\beta)}$. In hot Shakura–Sunyaev discs, $\beta = 3/7$ (Eq. 1.81), and in irradiation dominated discs, $\beta = 1/3$ (Eq. 1.92). During the first decay phase, the outer disc radius is almost constant so that using Eq. (1.97) the disc-mass evolution can be written as:

$$\frac{dM_d}{dt} = -\dot{M}_{in} \propto M_d^{1+\beta} \tag{1.98}$$

showing that $\dot{M}_{in}$ evolves almost exponentially, as long as $\dot{M}_{in}^{\beta}$ can be considered as constant (i.e. over about a decade in $\dot{M}_{in}$). "Exponential" decays in the DIM are only approximately exponential.

The quasi-exponential decay is due to *two* effects:

1. X-ray irradiation keeps the disc ionized, preventing cooling-front propagation;
2. tidal torques keep the outer disc radius roughly constant.

"Linear" Decay

The second phase of the decay begins when a disc ring cannot remain in thermal equilibrium. Locally, this corresponds to a fall onto the cool branch of the S-curve. In an irradiated disc, this happens when the central object does not produce enough

X-ray flux to keep the $T_{irr}(R_{out})$ above $\sim 10^4$ K. A cooling front appears and propagates down the disc at a speed of $v_{front} \approx \alpha_h c_s$.

In an irradiated disc, however, the transition between the hot and cold regions is set by T_{irr} because a cold branch exists only for $T_{irr} \lesssim 10^4$ K. In an irradiated disc, a cooling front can propagate inwards only down to the radius at which $T_{irr} \approx 10^4$ K, i.e. as far as there is a cold branch to fall onto. Thus, the decay is still irradiation-controlled. The hot region remains close to steady state, but its size shrinks $R_{hot} \sim \dot{M}_{in}^{1/2}$ (as shown in Eq. 1.92 with $T_{irr}(R_{hot}) = $ const).

Thermal Decay

In the model shown in Fig. 1.8, irradiation is unimportant after $t \gtrsim 170 - 190$ days because η becomes very small for $\dot{M}_{in} < 10^{16}$ g·s^{-1} when an ADAF forms. The cooling front thereafter propagates freely inwards, on a thermal timescale. In this particular case, the decrease of irradiation is caused by the onset of evaporation at the inner edge which lowers the efficiency. In general, there is always a moment at which T_{irr} becomes less than 10^4 K; evaporation just shortens the "linear" decay phase.

1.4.4 Maximum Accretion Rate and Decay Timescale

Now, we will see that there are two observable properties of X-ray transients that, when related one to the other, provide information and constraints on the physical properties of the outbursting system. The first is the maximum accretion rate $\dot{M}_{max}$ (Eq. 1.96). The second is the decay time of the X-ray flux: as we have seen, disc irradiation by the central X-rays traps the disc in the hot, high state, and only allows a decay of $\dot{M}$ on the hot-state viscous timescale. This is

$$t \simeq \frac{R^2}{3\nu} \tag{1.99}$$

which using Eq. (1.9) gives

$$t \simeq \frac{(GMR)^{1/2}}{3\alpha c_s^2}. \tag{1.100}$$

Taking the critical midplane temperature $T_c^+ \approx 16000$ K, one gets for the decay timescale

$$t \approx 32\, m^{1/2} R_{d,11}^{1/2} \alpha_{0.2}^{-1} \text{ days}, \tag{1.101}$$

where $\alpha_{0.2} = \alpha/0.2$. Eliminating R between (1.96) and (1.101) gives the accretion rate through the disc at the start of the outburst as

$$\dot{M} = 5.4 \times 10^{17}\, m^{-3.03}\, (t_{30}\alpha_{0.2})^{4.78} \text{ g s}^{-1}, \tag{1.102}$$

with $t = 30\,t_{30}$ d. Assuming an efficiency of η of 10%, the corresponding luminosity is

$$L = 5.0 \times 10^{37}\,\eta_{0.1} m^{-3.03}\,(t_{30}\alpha_{0.2})^{4.78}\ \text{erg s}^{-1}. \tag{1.103}$$

1.4.5 Comparison with Observations

1.4.5.1 Sub-Eddington Outbursts

The peak luminosities of most of the soft X-ray transients are sub-Eddington. Equation (1.102) can be written using the Eddington ratio $m := \dot{M}/\dot{M}_{\text{Edd}}$ as

$$\dot{m} = 0.42\eta_{0.1}(\alpha_{0.2}t_{30})^{4.78}m^{-4.03}. \tag{1.104}$$

This equation shows that the outburst peak will be sub-Eddington only if the outburst decay time is relatively short or the accretor (black hole) mass is high; i.e. the observed decay timescale is

$$t \lesssim 50\,\eta_{0.1}^{-0.21}\alpha_{0.2}^{-1}m^{0.84}\ \text{d}, \tag{1.105}$$

in good agreement with the compilation of X-ray transients outburst durations found in [59]. This shows that the standard value of efficiency $\eta_{0.1} \simeq 1$, and the value $\alpha_{0.2} \simeq 1$ deduced from observations of dwarf novae, gives the correct order of magnitude for the decay timescale of X-ray transients (from $\approx$3 days to $\approx$300 days). This equation also implies that black hole transients should have longer decay timescales than neutron star transients, all else being equal. Yan and Yu [59] find that outbursts last on average $\approx$2.5 $\times$ longer in black hole transients than in neutron star transients thus confirming this conclusion.

For sub-Eddington outbursts, Eq. (1.103) gives a useful relationship between distance D, bolometric flux F and outburst decay time t,

$$D_{\text{Mpc}} \simeq 1.0\,m^{-1.5}\left(\frac{\eta_{0.1}}{F_{12}}\right)^{1/2}(\alpha_{0.2}t_{50})^{2.4}, \tag{1.106}$$

where $D = D_{\text{Mpc}}$ Mpc and $F = 10^{-12}F_{12}$ erg s^{-1} cm^{-2}; $F = L/4\pi D^2$ and $t = 50\,t_{50}$ d.

Equation (1.106) shows that distant ($D > 1$Mpc) X-ray sources exhibiting variability typical of soft X-ray transients cannot contain black holes with masses superior to stellar masses [33].

1.4.5.2 Observational Tests

Finally, one can test observationally if soft X-ray transients satisfy the necessary condition for instability $\dot{M}_{\text{tr}} < \dot{M}_{\text{crit}}(R_d)$, where $\dot{M}_{\text{crit}}$ is the critical accretion rate for either non-irradiated or irradiated discs. In Fig. 1.9, the critical accretion rates (1.90) and (1.95) for respectively non-irradiated and irradiated disc around black holes are plotted as $\dot{M}(P_{\mathrm{orb}})$ relation. This relation was obtained from disc radius—orbital separation relation $R_d(a)$ [42], where (from Kepler's law) the orbital separation $a = 3.53 \times 10^{10}(m_1 + m_2)^{1/3}P_{\mathrm{hr}}^{2/3}$ cm, where m_i are the masses of the components in solar units, and P_{hr} is the orbital period in hours. Against these two critical lines, the actual positions of the observed sources are marked. The mass-transfer rate being difficult to measure, a proxy in the form of the *accumulation* rate

$$\dot{M}_{\mathrm{accum}} = \frac{\Delta E}{t_{\mathrm{rec}}\eta c^2} \qquad (1.107)$$

has been used. ΔE is the energy corresponding to the integrated X-ray luminosity from during an outburst and t_{rec} the recurrence time of the outbursts. One can see that all low-mass-X-ray-binary (LMXB) transients are in the unstable part of the figure, as they should be if the model is correct. One can also see that all

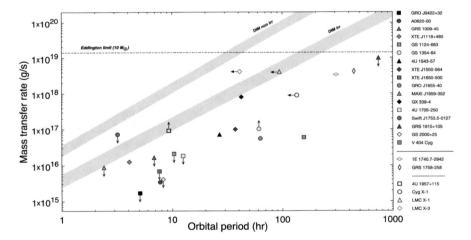

Fig. 1.9 Mass-transfer rate as a function of the orbital period for SXTs with *black holes*. The transient and persistent sources have been marked with respectively *filled* and *open symbols*. The *shaded grey areas* indicated "DIM irr" and "DIM non irr" represent the separation between persistent (*above*) and transient systems (*below*) according to the disc instability model when, respectively, irradiation is taken into account and when it is neglected. The horizontal *dashed line* indicates the Eddington accretion rate for a 10 $M_\odot$ black hole. All the upper limits on the mass-transfer rate are due to lower limits on the recurrence time. The upper limits on the mass-transfer rate of 4U 1957+115 and GS 1354-64 result from lower limits on the distance to the sources. The three left closed arrows do not indicate actual upper limits on the orbital period of Cyg X-1, LMC X-1 and LMC X-3. They emphasize that the radius of any accretion disc in these three high-mass XRBs is likely to be smaller than the one derived from the orbital period since they likely transfer mass by a (possibly focused) stellar wind instead of fully developed Roche lobe overflow. In the legend, the *solid horizontal line* separates transient and persistent systems. (The *dashed horizontal line* stresses that the persistent nature of 1E 1740.7-2942 and GRS 1758-258 is unclear.) [From [13]]

black hole LMXBs are transient. This is not true of neutron star LMXBs. Cyg X-1 in which the stellar companion of the black hole is a massive star is observed to be stable but according to Fig. 1.9 should be transient. This is not a problem because in such a system matter from the high-mass companion is not transferred by Roche-lobe overflow as in LMXBs, but lost through a stellar wind. In this case, the $R_d(a)$ relation used in the plot is not valid—the discs in such systems are smaller which is marked by a left-directed arrow at the symbol marking the position of this and two other similar objects (LMC X-1 and LMC X-3).

Additional reading: References [15, 16, 21, 31, 32].

1.5   Black Holes and Advection of Energy

Until now, we have neglected advection terms in the energy and momentum equations for stationary accretion flows. There two regimes of parameters where this assumption is not valid, in both cases for the same reason: low radiative efficiency when the time for radial motion towards the black hole is shorter than the radiative cooling time. Low density (low accretion rate), hot, optically thin accretion flows are poor coolers, and they are one of the two configurations where advection instead of radiation is the dominant evacuation-of-energy ("cooling") mechanism. Such optically thin flows are called ADAFs, for advection-dominated accretion flows. Also, advection-dominated are high-luminosity flows accreting at high rates, but they are called "slim discs" to account for their property of not being thin but still being described as if this were not of much importance.

    We shall start with optically thin flows.

- ADAFs
  Advection-dominated accretion flow (ADAF) is a term describing accretion of matter with angular momentum, in which radiation efficiency is very low. In their applications, ADAFs are supposed to describe inflows onto compact bodies, such as black holes or neutron stars; but very hot, optically thin flows are bad radiators in general so that, in principle, ADAFs are possible in other contexts. Of course in the vicinity of black holes or neutron stars, the virial (gravitational) temperature is $T_{\rm vir} \approx 5 \times 10^{12} (R_S/R)$ K, so that in optically thin plasmas, at such temperatures, both the coupling between ions and electrons and the efficiency of radiation processes are rather feeble. In such a situation, the thermal energy released in the flow by the viscosity, which drives accretion by removing angular momentum, is not going to be radiated away, but will be *advected* towards the compact body. If this compact body is a black hole, the heat will be lost forever, so that advection, in this case, acts as sort of a "global" cooling mechanism. In the case of infall onto a neutron star, the accreting matter lands on the star's surface and the (reprocessed) advected energy will be radiated away. There, advection may act only as a "local" cooling mechanism. (One should keep in mind that, in general,

advection may also be responsible for heating, depending on the sign of the temperature gradient—in some conditions, near the black hole, advection heats up electrons in a two-temperature ADAF.)

In general, the role of advection in an accretion flow depends on the radiation efficiency which in turns depends on the microscopic state of matter and on the absence or presence of a magnetic field. If, for a given accretion rate, radiative cooling is not efficient, advection is necessarily dominant, assuming that a stationary solution is possible.

- Slim discs

At high accretion rates, discs around black holes become dominated by radiation pressure in their inner regions, close to the black hole. At the same time, the opacity is dominated by electron scattering. In such discs, H/R is no longer $\ll 1$. But this means that terms involving the radial velocity are no longer negligible since $v_r \sim \alpha c_s(H/R)$. In particular, the advective term in the energy conservation equation $v_r \partial S/\partial R$ (see Eq. 1.39) becomes important and finally, at super-Eddington rates, dominant. When $Q^+ = Q^{\mathrm{adv}}$, the accretion flow is advection dominated and called a slim disc.

1.5.1 Advection-Dominated-Accretion-Flow Toy Models

One can illustrate the fundamental properties of ADAFs and slim discs with a simple toy model. The advection "cooling" (per unit surface) term in the energy equation can be written as

$$Q^{\mathrm{adv}} = \frac{\dot{M}}{2\pi R^2} c_s^2 \xi_a \tag{1.108}$$

(see Eq. 1.239).

Using the (nonrelativistic) hydrostatic equilibrium equation

$$\frac{H}{R} \approx \frac{c_s}{v_K} \tag{1.109}$$

one can write the advection term as

$$Q^{\mathrm{adv}} = \Upsilon \frac{\kappa_{\mathrm{es}} c}{2R} \left(\frac{\dot{m}}{\eta}\right) \xi_a \left(\frac{H}{R}\right)^2 \tag{1.110}$$

whereas the viscous heating term can be written as

$$Q^+ = \Upsilon \frac{3}{8} \frac{\kappa_{\mathrm{es}} c}{R} \left(\frac{\dot{m}}{\eta}\right), \tag{1.111}$$

where

$$\Upsilon = \left(\frac{cR_S}{\kappa_{es}R}\right)^2. \tag{1.112}$$

Since $\xi_a \sim 1$,

$$Q^{adv} \approx Q^+ \left(\frac{H}{R}\right)^2 \tag{1.113}$$

and, as said before, for geometrically thin discs ($H/R \ll 1$), the advective term Q^{adv} is negligible compared to the heating term Q^+ and in thermal equilibrium viscous heating must be compensated by radiative cooling. Things are different at, very high temperatures, when $(H/R) \sim 1$. Then, the advection term is comparable to the viscous term and cannot be neglected in the equation of thermal equilibrium. In some cases, this term is larger than the radiative cooling term Q^- and (most of) the heat released by viscosity is *advected* toward the accreting body instead of being locally radiated away as happens in geometrically thin discs.

From Eq. (1.57), one can obtain a useful expression for the square of the relative disc height (or aspect ratio):

$$\left(\frac{H}{R}\right)^2 = \frac{\sqrt{2}}{\kappa_{es}}\left(\frac{\dot{m}}{\eta}\right)(\alpha\Sigma)^{-1}\left(\frac{R_S}{R}\right)^{1/2}. \tag{1.114}$$

Deriving Eq. (1.114), we used the viscosity prescription $\nu = (2/3)\alpha c_s^2/\Omega_K$.

Using this equation, one can write for the advective cooling

$$Q^{adv} = \Upsilon\Omega_K\xi_a(\alpha\Sigma)^{-1}\left(\frac{\dot{m}}{\eta}\right)^2. \tag{1.115}$$

The thermal equilibrium (energy) equation is

$$Q^+ = Q^{adv} + Q^-. \tag{1.116}$$

The form of the radiative cooling term depends on the state of the accreting matter, i.e. on its temperature, density and chemical composition. Let us consider two cases of accretion flows:

- optically thick
  and
- optically thin.

For the optically thick case, we will use the diffusion approximation formula

$$Q^- = \frac{8}{3}\frac{\sigma T_c^4}{\kappa_R\Sigma}, \tag{1.117}$$

and assume $\kappa_R = \kappa_{es}$. With the help of Eq. (1.114), this can be brought to the form

$$Q_{\text{thick}}^- = 8\Upsilon \left(\frac{\kappa_{es}R_S}{c}\right)^{1/2} \left(\frac{R}{R_S}\right)^2 \Omega_K^{3/2} (\alpha\Sigma)^{-1/2} \left(\frac{\dot{m}}{\eta}\right)^{1/2}. \tag{1.118}$$

For the optical thin case of bremsstrahlung radiation, we have

$$Q^- = 1.24 \times 10^{21} H\rho^2 T^{1/2} \tag{1.119}$$

which using Eq. (1.114) can be written

$$Q_{\text{thin}}^- = 3.4 \times 10^{-6}\Upsilon \left(\frac{R}{R_S}\right)^2 \Omega_K \alpha^{-2} (\alpha\Sigma)^2. \tag{1.120}$$

- In the OPTICALLY THICK case, we have therefore

$$\xi_a \left(\frac{\dot{m}}{\eta}\right)^2 + 0.18 \left(\frac{R}{R_S}\right)^{1/2} (\alpha\Sigma) \left(\frac{\dot{m}}{\eta}\right)$$
$$+ 2.3 \left(\frac{R}{R_S}\right)^{5/4} (\alpha\Sigma)^{1/2} \left(\frac{\dot{m}}{\eta}\right)^{1/2} = 0 \tag{1.121}$$

- In the OPTICALLY THIN case, the energy equation has the form

$$\xi_a \left(\frac{\dot{m}}{\eta}\right)^2 + 0.18 \left(\frac{R}{R_S}\right)^{1/2} (\alpha\Sigma) \left(\frac{\dot{m}}{\eta}\right)$$
$$+ 3 \times 10^{-6}\alpha^{-2} \left(\frac{R}{R_S}\right)^2 (\alpha\Sigma)^3 = 0 \tag{1.122}$$

There are two distinct types of advection-dominated accretion flows: optically thin and optically thick. We will first deal with optically thin flows known as *ADAFs*.

1.5.1.1   Optically Thin Flows: *ADAFs*

For prescribed values α and ξ_a, Eq. (1.122) is a quadratic equation in $(\dot{m}/\eta)$ whose solutions in the form of $\dot{m}(\Sigma)$ describe thermal equilibria at a given value of R. Obviously, for a given Σ, this equation has at most two solutions. The solutions form two branches on the $\dot{m}(\alpha\Sigma)$—plane:

- the ADAF branch

$$\dot{m} = 0.53\kappa_{es}\,\eta \left(\frac{R}{R_S}\right)^{1/2} \xi_a^{-1}\alpha\Sigma. \tag{1.123}$$

and
• the radiatively cooled branch

$$\dot{m} = 1.9 \times 10^{-5} \, \eta \left(\frac{R}{R_S}\right)^{3/2} \xi_a^{-1} \alpha^{-2} \, (\alpha \Sigma)^2 \, . \tag{1.124}$$

From Eqs. (1.123 and 1.124), it is clear that there exists a maximum accretion rate for which only one solution of Eq. (1.122) exists. This implies the existence of a maximum accretion rate at

$$\dot{m}_{\text{max}} \approx 1.7 \times 10^3 \eta \, \alpha^2 \left(\frac{R}{R_S}\right)^{1/2} \, . \tag{1.125}$$

This is where the two branches formed by thermal equilibrium solutions on the $\dot{m}(\alpha \Sigma)$—plane meet as seen on Fig. 1.10.

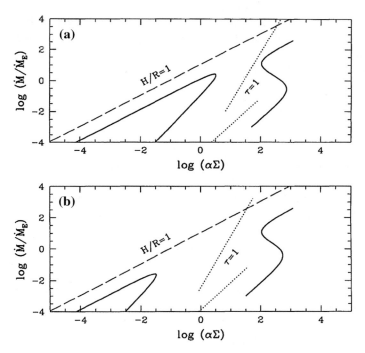

Fig. 1.10 a Thermal equilibria for optically thick (The *right solid S-shaped line*) and optically thin (the *left solid line*) accretion flows. The upper branches represent advection-dominated solution (ADAFs). Flows above the *dotted lines* $\tau = 1$ are optically thin $-\,-\,-\tau$ is the effective optical depth calculated for radiation-pressure-dominated (*upper line*) or gas-dominated (*lower line*) configurations. It is assumed that $M_{\text{BH}} = 10 \, M_\odot$, $R = 5R_S$, $\alpha = 0.1$ and $\xi_a = 1$. **b** The same for $\alpha = 0.01$ [From [4]]

The value of $\dot{m}_{\max}$ depends on the cooling mechanism in the accretion flow; the free–free cooling is not a realistic description of the emission in the vicinity ($(R/R_S) \lesssim 10^3$) of a black hole. The flow there most probably forms a two-temperature plasma. In such a case, $\dot{m}_{\max} \approx 10\alpha^2$ with almost no dependence on radius. For larger radii, $\dot{m}_{\max}$ decreases with radius.

1.5.1.2   Optically Thick Flows: *Slim Discs*

Since the first two terms in Eq. (1.121) are the same as in (Eq. 1.122), the high, $\dot{m}$, advection-dominated solution is the same as in the optically thin case but now represents the

- Slim-disc branch

$$\dot{m} = 0.53\,\kappa_{\mathrm{es}}\,\eta\left(\frac{R}{R_S}\right)^{1/2}\xi_a^{-1}\alpha\Sigma. \tag{1.123}$$

Now, the full Eq. (1.121) is a cubic equation in $\dot{m}^{1/2}$, and on the $\dot{m}(\alpha\Sigma)$ plane, its solution forms the two upper branches of the S-curve shown in Fig. 1.10. The uppermost branch corresponds to slim discs, while the branch with negative slope represents the Shakura–Sunayev solution in the regime a. (see Sect. 1.3.6), i.e.

- a radiatively cooled, radiation-pressure-dominated accretion disc

$$\dot{m} = 160\,\kappa_{\mathrm{es}}^{-1}\,\eta\left(\frac{R}{R_S}\right)^{3/2}(\alpha\Sigma)^{-1} \tag{1.126}$$

1.5.1.3   Thermal Instability of Radiation–Pressure-Dominated Discs

Radiation-pressure-dominated ($P = P_{\mathrm{rad}}$) accretion discs are thermally unstable when opacity is due to electron scattering on electrons. Indeed,

$$\frac{d\ln T_{\mathrm{eff}}^4}{d\ln T_{\mathrm{c}}} = 4 \tag{1.127}$$

because $\kappa_R = \kappa_{\mathrm{es}} = const.$, while in a radiation pressure-dominated disc $Q^+ \sim \nu\Sigma \sim HT^4 \sim T^8/\Sigma$ so

$$\frac{d\ln Q^+}{d\ln T_{\mathrm{c}}} = 8 > \frac{d\ln T_{\mathrm{eff}}^4}{d\ln T_{\mathrm{c}}} \tag{1.128}$$

and the disc is thermally unstable. This solution is represented by the middle branch with negative slope (see Eq. 1.126) in Fig. 1.10. The presence of this instability in the model is one of the unsolved problems of the accretion disc theory because it contradicts observations which do not show any unstable behaviour in the range of luminosities where discs should be in the radiative pressure and electron-scattering opacity domination regime.

1.5.1.4   Slim Discs and Super-Eddington Accretion

From Eqs. (1.114) and (1.126), one obtains for the disc aspect ratio

$$\frac{H}{R} = 0.11 \left(\frac{\dot{m}}{\eta}\right) \frac{R_S}{R} \qquad (1.129)$$

which shows that the height of a radiation dominated disc is constant with radius and proportional to the accretion rates.

But this means that with increasing $\dot{m}$, advection becomes more and more important (see, e.g. Eq. 1.113) and for

$$\frac{\dot{m}}{\eta} \approx 9.2 \frac{R}{R_S} \qquad (1.130)$$

advection will take over radiation as the dominant cooling mechanism and the solution will represent a slim disc. Equation (1.130) can be also interpreted as giving the *transition radius* between radiatively and advectively cooled disc for a given accretion rate $\dot{m}$:

$$\frac{R_{\text{trans}}}{R_S} \approx \frac{0.1}{\eta} \dot{m} \qquad (1.131)$$

Another radius of interest is the *trapping radius* at which the photon diffusion (escape) time $H\tau/c$ is equal to the viscous infall time R/v_r

$$R_{\text{trapp}} = \frac{H\tau v_r}{c} = \frac{H\kappa\Sigma}{c} \frac{\dot{M}}{2\pi R\Sigma} = \frac{H}{R}\left(\frac{\dot{m}}{\eta}\right) R_S. \qquad (1.132)$$

Notice that both R_{trans} and R_{trapp} are proportional to the accretion rate.

In an advection-dominated disc, the aspect ration H/R is independent of the accretion rate:

$$\frac{H}{R} = 0.86\,\xi_a \left(\frac{R}{R_S}\right)^{1/4}, \qquad (1.133)$$

and therefore contrary to radiatively cooled discs, slim disc do not puff up with increasing accretion rate.

Putting (1.133) into Eq. (1.132), one obtains

$$\frac{R_{\text{trapp}}}{R} = 0.86\,\xi_a^{-1/2} \left(\frac{R}{R_S}\right)^{1/4} \left(\frac{\dot{m}}{\eta}\right). \qquad (1.134)$$

Radiation inside the trapping radius is unable to stop accretion, and since $R_{\text{trapp}} \sim \dot{m}$, there is no limit on the accretion rate onto a black hole.

The luminosity of the toy-model slim disc can be calculated from Eqs. (1.118) and (1.123) giving

$$Q^- = \sigma T_{\text{eff}}^4 = \frac{0.1}{\xi_a} \frac{L_{\text{Edd}}}{R^2}, \qquad (1.135)$$

which implies $T_{\text{eff}} \sim 1/R^{1/2}$. The luminosity of the slim-disc part of the accretion flow is then

$$L_{\text{slim}} = 2 \int_{R_{\text{in}}}^{R_{\text{trans}}} \sigma T_{\text{eff}}^4 2\pi R dR = \frac{0.8}{\xi_a} L_{\text{Edd}} \cdot \ln \frac{R_{\text{trans}}}{R_{\text{in}}} \approx L_{\text{Edd}} \ln \dot{m}, \qquad (1.136)$$

where we used Eq. (1.131).

Therefore, the total disc luminosity

$$L_{\text{total}} = L_{\text{thin}} + L_{\text{slim}}$$
$$= 4\pi \left(\int_{R_{\text{in}}}^{R_{\text{trans}}} \sigma T_{\text{eff}}^4 R dR + \int_{R_{\text{trans}}}^{R_\infty} \sigma T_{\text{eff}}^4 R dR \right) \approx L_{\text{Edd}} (1 + \ln \dot{m}), \quad (1.137)$$

where L_{thin} is the luminosity of the radiation-cooled disc for which Eq. (1.58) applies.

It is easy to see that the same luminosity formula $L \approx L_{\text{Edd}}(1 + \ln \dot{m})$ is obtained when one assumes mass loss from the disc resulting in a variable (with radius) accretion rate: $\dot{M} \sim R$.

At very high accretion rates, the disc emission will be also strongly beamed by the flow geometry so that observer situated in the beam of the emitting system will infer a luminosity

$$L_{\text{sph}} = \frac{1}{b} L_{\text{Edd}} (1 + \ln \dot{m}), \qquad (1.138)$$

where b is the beaming factor (see [27] for a derivation of b in the case of ultraluminous X-ray sources).

Numerical simulations do not seem to correspond to this analytical solutions (see e.g. [25, 51, 52], but they also disagree between themselves. The reasons for these contradictions are worth investigating.

Additional reading: References [1, 3, 4, 33, 38, 49, 50, 60].

1.6   Accretion Discs in Kerr Spacetime

In this section, we will present and discuss the set equations whose solutions represent α-accretion discs in the Kerr metric. This section is based on references [5, 30, 49] and to be understood requires some basic knowledge of Einstein's general relativity.

1.6.1   Kerr Black Holes

The components g_{ij} of the metric tensor with respect to the coordinates (t, x^α) are expressible in terms of the lapse N, the components β^α of the shift vector and the components $\gamma_{\alpha\beta}$ of the spatial metric:

$$g_{ij}\, dx^i\, dx^j = -N^2 dt^2 + \gamma_{\alpha\beta}(dx^\alpha + \beta^\alpha dt)(dx^\beta + \beta^\beta dt), \qquad (1.139)$$

which is a modern way of writing the metric.

Remark 1.3 In this section only I will use conventions different from those used in other parts of the chapter. First, I will use the so-called *geometrical units* that are linked to the physical units for length, time and mass by

$$\text{length in physical units} = \text{length in geometrical units},$$
$$\text{time in physical units} = \frac{1}{c}\,\text{length in geometrical units},$$
$$\text{mass in physical units} = \frac{c^2}{G}\,\text{length in geometrical units}.$$

$$(1.140)$$

Second, the radial coordinate will be called "r" and not "R". This should not confuse the reader since R is used only in the *nonrelativistic* context where it denotes a radial coordinate and a radial distance, while in the relativistic context, it only a coordinate.

1.6.1.1 General Structure, Boyer–Lindquist Coordinates

The Kerr metric in the Boyer–Lindquist (spherical) coordinates t, r, θ, φ corresponds to:

$$N = \frac{\varsigma}{\sqrt{A\Delta}}, \quad \beta^r = \beta^\theta = 0, \quad \beta^\varphi = -\omega, \qquad (1.141)$$

$$g_{rr} = \frac{\varsigma^2}{\Delta}, \quad g_{\theta\theta} = \varsigma^2, \quad g_{\varphi\varphi} = \frac{A^2}{\varsigma^2}\sin^2\theta \qquad (1.142)$$

with

$$\varsigma = r^2 + a^2 \cos^2\theta, \quad \Delta = r^2 - 2Mr + a^2, \qquad (1.143)$$

$$A = \left(r^2 + a^2\right)^2 - \Delta a^2 \sin^2\theta, \quad \omega = \frac{2Jr}{A} = \frac{2Mar}{A}, \qquad (1.144)$$

where M is the mass and $a = J/M$ is the angular momentum per unit mass. In application,s one often uses the dimensionless "angular momentum" parameter $a_* = a/M$.

Therefore in BL coordinates, the Kerr metric takes the form of

$$ds^2 = -\frac{\varsigma^2 \Delta}{A}dt^2 + \frac{A\sin^2\theta}{\varsigma^2}(d\varphi - \omega dt)^2 + \frac{\varsigma^2}{\Delta}dr^2 + \varsigma^2 d\theta^2. \qquad (1.145)$$

The time (stationarity) and axial symmetries of the metric are expressed by two Killing vectors

$$\eta^i = \delta^i{}_{(t)}, \quad \xi^i = \delta^i{}_{(\varphi)}, \qquad (1.146)$$

where $\delta^i{}_{(k)}$ is the Kronecker delta.

Remark 1.4 Using Killing vectors (1.146), one can define some useful scalar functions: the angular velocity of the dragging of inertial frames ω, the gravitational potential Φ and the gyration radius $\mathfrak{R}$,

$$\omega = -\frac{\eta \cdot \xi}{\xi \cdot \xi}, \quad e^{-2\Phi} = \omega^2 \xi \cdot \xi - \eta \cdot \eta, \quad \mathfrak{R}^2 = -\frac{\xi \cdot \xi}{\eta \cdot \eta}. \qquad (1.147)$$

In the Boyer–Lindquist coordinates, the scalar products of the Killing vectors are simply given by the components of the metric,

$$\eta \cdot \eta = g_{tt}, \quad \eta \cdot \xi = g_{t\varphi}, \quad \xi \cdot \xi = g_{\varphi\varphi}, \qquad (1.148)$$

and therefore, quantities defined in Eq. (1.147) can be explicitly written down in terms of the Boyer–Lindquist coordinates as:

$$\mathfrak{R}^2 = \frac{A^2}{r^4 \Delta}, \quad e^{-2\Phi} = \frac{r^2 \Delta}{A}. \qquad (1.149)$$

• **The horizon**

The black hole surface (event horizon) is at

$$r_H = M + \sqrt{M^2 - a^2}. \qquad (1.150)$$

Therefore, a horizon exists for $a_* \le 1$ only. At the horizon, the angular velocity of the dragging of inertial frame is equal to

$$\omega_H = \Omega_H = \frac{a}{2Mr_H}, \qquad (1.151)$$

where Ω_H is the angular velocity of the horizon, i.e. the angular velocity of the horizon-forming light rays with respect to infinity. The horizon rotates.
The area of the horizon is given by

$$S = 8\pi M r_H = 8\pi M \sqrt{M^2 - a^2}. \qquad (1.152)$$

The extreme (maximally rotating) black hole corresponds to

$$a = M. \tag{1.153}$$

For $a > M$, the Kerr solution represents a *naked singularity*. Such singularities would be a great embarrassment not only because of their visibility but also because the solution of Einstein equation in which they appear violate causality by containing closed timelike lines. The conjecture that no naked singularity is formed through collapse of real bodies is called the *cosmic censorship hypothesis* (Roger Penrose).

Remark 1.5 **Rotation of astrophysical bodies**
Since this is a lecture in astrophysics, let us leave for a moment the geometrical units. They are great for calculations but usually useless for comparing their results with observations. In the physical units,

$$r_H = \frac{GM}{c^2} + \left[\left(\frac{GM}{c^2} \right)^2 - \left(\frac{J}{Mc} \right)^2 \right]^{1/2}. \tag{1.154}$$

Therefore, the maximum angular momentum of a black hole is

$$J_{max} = \frac{GM^2}{c} = 8.9 \times 10^{48} \left(\frac{M}{M_\odot} \right)^2 \text{ g cm}^2 \, s^{-1}. \tag{1.155}$$

This is slightly more than the angular momentum of the Sun ($J_\odot = 1.63 \times 10^{48}$ g cm^2 s^{-1}, $a_*^\odot = 0.185$): the gain in velocity is almost fully compensated by the loss in radius.

For a millisecond pulsar which is a neutron star with a mass of $\sim 1.4 \, M_\odot$ and radius $\sim$10km, the angular momentum is

$$J_{NS} = I_{NS} \Omega_S \approx 8.6 \times 10^{48} \left(\frac{\alpha(x)}{0.489} \right) \left(\frac{M_{NS}}{1.4 \, M_\odot} \right) \left(\frac{R_{NS}}{10 \text{ km}} \right)^2 \left(\frac{P_S}{1 \text{ ms}} \right)^{-1} \text{ g cm}^2 \tag{1.156}$$

where $I_{NS} \approx \alpha(x) M_{NS} R_{NS}^2$ is the moment of inertia and $x = (M_{NS}/ M_\odot)(km/R_{NS})$ the compactness parameter. For the most compact neutron star, $x \leq 0.24$ and $\alpha(x) \lesssim 0.489$. Therefore for neutron stars that rotate at millisecond periods

$$a_*^{NS} \approx 0.5 \left(\frac{\alpha(x)}{0.489} \right) \left(\frac{M_{NS}}{1.4 \, M_\odot} \right)^{-1} \left(\frac{R_{NS}}{10 \text{ km}} \right)^2 \left(\frac{P_S}{1 \text{ ms}} \right)^{-1}. \tag{1.157}$$

By definition

- the specific (per unit rest-mass) energy is

$$\mathfrak{E} := -\boldsymbol{\eta} \cdot \mathbf{u}, \tag{1.158}$$

- the specific (per unit rest-mass) angular momentum is

$$\mathfrak{L} := \boldsymbol{\xi} \cdot \mathbf{u} \tag{1.159}$$

and

- the specific (per unit mass-energy) angular momentum (also called *geometrical specific angular momentum*) is

$$\mathfrak{J} := -\frac{\mathfrak{L}}{\mathfrak{E}} = -\frac{\boldsymbol{\xi} \cdot \mathbf{u}}{\boldsymbol{\eta} \cdot \mathbf{u}} \tag{1.160}$$

1.6.2 Privileged Observers

Let us consider observers privileged by the symmetries of the Kerr spacetime. The results below apply to any spacetime with the same symmetries, e.g. the spacetime of a stationary, rotating star. The four-velocity of a privileged observer is the linear combination of the two Killing vectors:

$$\mathbf{u} = Z \left(\boldsymbol{\eta} + \Omega_{\mathrm{obs}} \boldsymbol{\xi} \right) \tag{1.161}$$

where the redshift factor Z is (from the normalization $\mathbf{u} \cdot \mathbf{u} = 1$)

$$Z^{-2} = \boldsymbol{\eta} \cdot \boldsymbol{\eta} + 2\Omega_{\mathrm{obs}} \boldsymbol{\eta} \cdot \boldsymbol{\xi} + \Omega_{\mathrm{obs}}^2 \boldsymbol{\xi} \cdot \boldsymbol{\xi} \tag{1.162}$$

Since for $a \neq 0$, the Kerr spacetime is stationary but not static; i.e. the timelike Killing vector η is not orthogonal to the spacelike surfaces $t =$const. In such a spacetime,"non-rotation" is not uniquely defined.

Stationary observers are immobile with respect to infinity; their four-velocities are defined as

$$u_{\mathrm{stat}}^i = (\eta\eta)^{-1/2} \eta^i \tag{1.163}$$

but are locally rotating: $\mathscr{L}_{\mathrm{stat}} = \xi_i u_{\mathrm{stat}}^i \neq 0$.

The four-velocity of a locally non-rotating observer is a unit timelike vector orthogonal to the spacelike surfaces $t =$const.:

$$u_{\mathrm{ZAMO}}^i = e^{\Phi} \left(\eta^i + \omega \xi^i \right), \tag{1.164}$$

Table 1.1 Summary of properties of privileged observers

Observer	Four-velocity	Angular velocity with respect to stationary observers
Stationary	$\mathbf{u} = (\boldsymbol{\eta} \cdot \boldsymbol{\eta})^{-1/2}\boldsymbol{\eta}$	$\Omega_{\text{stat}} = 0$
ZAMO (LNR)	$\mathbf{u} = e^{-\Phi}(\boldsymbol{\eta} + \omega\boldsymbol{\xi})$	$\Omega_{\text{ZAMO}} = \omega$
Comoving (with matter)	$\mathbf{u} = Z(\boldsymbol{\eta} + \Omega\boldsymbol{\xi})$	$\Omega_{\text{com}} = \Omega$

defines the four-velocity of the local inertial observer or ZAMO, i.e. zero-angular momentum observers since

$$\mathscr{L}_{\text{ZAMO}} = \xi_i u^i_{\text{ZAMO}} = 0.$$

Finally, in the presence of matter forming a stationary and axisymmetric configuration, there are privileged observers *comoving* with matter (Table 1.1).

1.6.3   The Ergosphere

For ZAMOs, $\Omega = \omega$, but for stationary observers, $\Omega - \omega = -\omega$. Therefore, ZAMOs rotate with respect to infinity (but are *locally* non-rotating). They may exist down to the black hole horizon, where they become null: $\mathbf{u}_{\text{ZAMO}} \cdot \mathbf{u}_{\text{ZAMO}} = 0$.

Stationary observers immobile with respect to infinity but rotating with angular velocity $-\omega$ with respect to ZAMOs can exist (their four-velocity must be timelike, $\boldsymbol{\eta} \cdot \boldsymbol{\eta} < 0$) only outside the *stationarity limit* whose radius is defined by $\boldsymbol{\eta} \cdot \boldsymbol{\eta} = 0$:

$$r_{\text{er}}(\theta) = M + \sqrt{M^2 - a^2\cos^2\theta}. \tag{1.165}$$

The stationarity limit is called the *ergosphere*.

1.6.4   Equatorial Plane

We will discuss now orbits in the equatorial plane, where they have the axial symmetry. We are introducing the cylindrical vertical coordinate $z = \cos\theta$ which is defined very close to the equatorial plane, $z = 0$. The metric of the Kerr black hole in the equatorial plane, accurate up to the $(z/r)^0$, is

$$ds^2 = -\frac{r^2\Delta}{A}dt^2 + \frac{A}{r^2}(d\varphi - \omega dt)^2 + r^2\frac{\Delta}{d}r^2 + dz^2, \tag{1.166}$$

where now

$$\Delta = r^2 - 2Mr + a^2, \quad A = \left(r^2 + a^2\right)^2 - \Delta a^2, \quad \omega = \frac{2Mar}{A}, \quad (1.167)$$

or simpler

$$ds^2 = -\left(1 - \frac{2M}{r}\right) dt^2 - 2\omega dt d\varphi + \frac{A}{r^2} d\varphi^2 + \frac{r^2}{\Delta} dr^2 + dz^2. \quad (1.168)$$

1.6.4.1   Orbits in the Equatorial Plane

The four-velocity of matter u^i has components u^t, u^φ, u^r,

$$u^i = u^t \delta^i_{\ (t)} + u^\varphi \delta^i_{\ (\varphi)} + u^r \delta^i_{\ (r)}. \quad (1.169)$$

The angular frequency Ω with respect to a stationary observer and the angular frequency $\tilde{\Omega}$ with respect to a local inertial observer are respectively defined by

$$\Omega = \frac{u^\varphi}{u^t}, \quad \tilde{\Omega} = \Omega - \omega, \quad (1.170)$$

The angular frequencies of the corotating $(+)$ and counterrotating $(-)$ *Keplerian orbits* are

$$\Omega_K^\pm = \pm \frac{M^{1/2}}{r^{3/2} \pm aM^{1/2}}, \quad (1.171)$$

the specific energy is

$$\mathfrak{E}_K^\pm = \frac{r^2 - 2Mr \pm a(Mr)^{1/2}}{r\left(r^2 - 3Mr \pm 2a(Mr)^{1/2}\right)^{1/2}}, \quad (1.172)$$

and the specific angular momentum is given by

$$\mathfrak{L}_K^\pm = \pm \frac{(Mr)^{1/2} \left(r^2 \mp 2a(Mr)^{1/2} + a^2\right)}{r\left(r^2 - 3Mr \pm 2a(Mr)^{1/2}\right)^{1/2}}, \quad (1.173)$$

or

$$\mathfrak{J}_K = \pm \frac{(Mr)^{1/2} \left(r^2 \mp 2a(Mr)^{1/2} + a^2\right)}{r^2 - 2Mr \pm a(Mr)^{1/2}}. \quad (1.174)$$

Both $\mathfrak{J}$ and $\mathfrak{L}$ have a minimum at the last stable orbit, more often called ISCO (innermost stable circular orbit).

Because of the rotation of space, there is no direct relation between angular momentum and angular frequency but

$$\mathfrak{J} = \frac{\eta \cdot \xi + \Omega \xi \cdot \xi}{\eta \cdot \eta + \Omega \eta \cdot \xi} = \frac{\mathfrak{R}^2 \Omega - \omega}{\Omega \omega - 1}. \tag{1.175}$$

For the Schwarzschild solution $(a = \omega = 0)$

$$\mathfrak{J}_K = \mathfrak{R}^2 \Omega_K, \tag{1.176}$$

so a Newtonian-like relation (justifying the name "gyration radius" for $\mathfrak{R}$) between angular frequency and angular momentum exists for $\mathfrak{J}$. No such relation exists for $\mathfrak{L}$.

ISCO

The minimum of the Keplerian angular momentum corresponding to the *innermost stable circular orbit* (ISCO) is located at

$$
\begin{aligned}
r_{\text{ISCO}}^{\pm} &= M\{3 + Z_2 \mp [(3 - Z_1)(3 + Z_1 + 2Z_2)]^{1/2}\}, \\
Z_1 &= 1 + \left(1 - a^2/M^2\right)^{1/3} \left[(1 + a/M)^{1/3} + (1 - a/M)^{1/3}\right], \\
Z_2 &= \left(3a^2/M^2 + Z_1^2\right)^{1/2}.
\end{aligned}
\tag{1.177}
$$

Binding Energy

The binding energy

$$\mathscr{E}_{\text{bind}} = 1 - \mathfrak{E}_K \tag{1.178}$$

at the ISCO is

- $1 - \sqrt{8/9} \approx 0.06$ for $a = 0$
- $1 - \sqrt{1/3} \approx 0.42$ for $a = 1$.

This corresponds to the efficiencies of accretion in a geometrically thin (quasi-Keplerian) disc around a black hole (Fig. 1.11).

For a Schwarzschild black hole, the frequency associated with the ISCO at $r_{\text{ISCO}} = 6M$ is

$$\nu_K(r_{\text{ISCO}}) = 2197 \left(\frac{M}{M_\odot}\right)^{-1} \text{Hz}. \tag{1.179}$$

IBCO

The binding energy of a Keplerian orbit $1 - \mathfrak{E} = 0$ at the marginally bound orbit (or IBCO: innermost bound circular orbit) is

$$r_{\text{IBCO}}^{\pm} = 2M \mp a + 2\sqrt{M^2 \mp aM}. \tag{1.180}$$

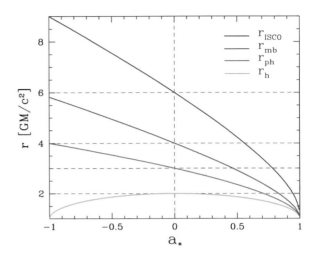

Fig. 1.11 Radii of characteristic orbits in the Kerr metric as a function of $a_* = a/M$. The innermost stable circular orbit: r_{ISCO}; the marginally bound orbit r_{IBCO} (marked r_{mb}); the photon orbit: r_{ph}; and the black hole horizon: r_H (marked r_h) (Courtesy of A. Sądowski)

For a non-rotating black hole, $r_{IBCO} = 4M$, and the frequency associated with the IBCO is

$$\nu_K(r_{IBCO}) = 4037 \left(\frac{M}{M_\odot}\right)^{-1} \text{Hz}. \tag{1.181}$$

ICO (*Circular photon orbit*)
The innermost circular orbit (ICO), i.e. the circular photon orbit, is at

$$r_{ph}^\pm = 2M \left(1 + \cos\left[\frac{2}{3}\cos^{-1}\left(\mp\frac{a}{M}\right)\right]\right). \tag{1.182}$$

For a non-rotating black hole, $r_{ph} = 3M$.

1.6.4.2 Epicyclic Frequencies

We will consider now consider a perturbed orbital motion in, and slightly off the equatorial plane. In the Newtonian case, the angular frequency of such motions must be equal to the Keplerian frequency Ω_K since there is only one characteristic scale defined by the gravitational constant G. In general relativity, the presence of two constants G and c implies that the epicyclic frequency does not have to be equal to Ω_K.

The four-velocity for the perturbed circular motion can be written as

$$u^i = \left(1, \tilde{u}^r, \tilde{u}^\theta, \Omega_K + \tilde{u}^\varphi\right), \tag{1.183}$$

where $\tilde{u}^\alpha$ are the velocity perturbations.

- For perturbations in the equatorial plane, the equation of motion is

$$\left(\frac{\partial^2}{\partial t^2} + \kappa^2\right)\left(\begin{matrix}\tilde{u}^r \\ \tilde{u}^\varphi\end{matrix}\right) = 0, \tag{1.184}$$

where

$$\kappa^2 = \Omega_K^2 \frac{r^2 - 6Mr \pm 8aM^{1/2}r^{1/2} - 3a^2}{r^2} \tag{1.185}$$

is the (equatorial) *epicyclic frequency*. In the Schwarzschild case $a = 0$ this is $\kappa^2 = \Omega_K^2(1 - 6M/r)$ and vanishes at ISCO. In the Newtonian limit, the epicyclic frequency equals to the Keplerian frequency $\kappa = \Omega_K$.
- For vertical perturbations, the equation is

$$\left(\frac{\partial}{\partial t} + \Omega_K \frac{\partial}{\partial \varphi}\right)\tilde{u}^\theta = -\Omega_\perp \delta\theta, \tag{1.186}$$

where the vertical epicyclic (angular) frequency is given by

$$\Omega_\perp^2 = \Omega_K^2 \frac{r^2 - 4aM^{1/2}r^{1/2} - 3a^2 M^2}{r^2} \tag{1.187}$$

In the Schwarzschild case ($a = 0$), the vertical epicyclic frequency is equal to the Keplerian angular frequency Ω_K, which is to be expected from the spherical symmetry of this solution.

The angular velocity $\Omega_\perp$ appears also in the equation of vertical equilibrium of a (quasi)Keplerian disc which will be discussed later (Sect. 1.8.5). Here, let us just notice that Eq. (1.222) can be written as

$$\frac{\partial p}{\partial z} = -\rho e^{2\Phi} \Omega_\perp^2 z. \tag{1.188}$$

All these characteristic frequencies can be put into the form

$$\Omega = f(x, a_*)\frac{1}{M}, \tag{1.189}$$

where $x = r/M$. For all relativistic frequencies, $x = x(a_*)$, and therefore, they can be written as

$$\Omega = F(a_*)\frac{1}{M}. \tag{1.190}$$

Additional reading: Reference [2].

1.7 Accretion Flows in the Kerr Spacetime

1.7.1 Kinematic Relations

In the reference frame of the local inertial (non-rotating) observer, the four-velocity takes the form,

$$u^i = \gamma \left(u^i_{\text{ZAMO}} + v^{(\varphi)} \tau^i_{\,(\varphi)} + v^{(r)} \tau^i_{\,(r)} \right). \tag{1.191}$$

The vectors $\tau^i_{\,(\varphi)}$ and $\tau^i_{\,(r)}$ are the unit vectors in the coordinate directions φ and r. The Lorentz gamma factor γ equals,

$$\gamma = \frac{1}{\sqrt{1 - \left(v^{(\varphi)}\right)^2 - \left(v^{(r)}\right)^2}}. \tag{1.192}$$

The relation between the Boyer–Lindquist and the physical velocity component in the azimuthal direction is,

$$v^{(\varphi)} = \tilde{R}\tilde{\Omega}, \tag{1.193}$$

which justifies the name of $\tilde{R}$—gyration radius. It is convenient to use the (rescaled) radial velocity component V defined by the formula,

$$\frac{V}{\sqrt{1 - V^2}} = \gamma v^{(r)} = u^r g_{rr}^{1/2}. \tag{1.194}$$

The Lorentz gamma factor may then be written as,

$$\gamma^2 = \left(\frac{1}{1 - \tilde{\Omega}^2 \tilde{R}^2} \right) \left(\frac{1}{1 - V^2} \right), \tag{1.195}$$

which allows writing a simple expression for V in terms of the velocity components measured in the frame of the local inertial observer,

$$V = \frac{v^{(r)}}{\sqrt{1 - \left(v^{(\varphi)}\right)^2}} = \frac{v^{(r)}}{\sqrt{1 - \tilde{R}^2 \tilde{\Omega}^2}}. \tag{1.196}$$

Thus, V is the radial velocity of the fluid as measured by an observer corotating with the fluid at fixed r.

Although a different quantity could have been chosen as the definition of the "radial velocity", only V has directly three very convenient properties, all guaranteed by its definition:

- (i) everywhere in the flow $|V| \leq 1$,
- (ii) on the horizon $|V| = 1$,

- (iii) at the sonic point $|V| \approx c_s$,

where c_s is the local sound speed.

To see that property (i) holds, let us define

$$\tilde{V}^2 = u^r u_r = u^r u^r g_{rr} \geq 0. \tag{1.197}$$

Then, one has

$$V^2 = \tilde{V}^2/(1 + \tilde{V}^2) \leq 1. \tag{1.198}$$

Writing $V = \sqrt{r^2 u^r u^r/(r^2 u^r u^r + \Delta)}$ demonstrates property (ii) since $|V| = 1$ independent of the value of $r^2 u^r u^r$.

For the proof of property (iii) of V see [4].

Other possible choices of the "radial velocity" such as $u = |u^r|$ are not that convenient.

1.7.2   Description of Accreting Matter

The stress–energy tensor T^{ik} of the matter in the disc is given by

$$T^{ik} = (\varepsilon + p)\, u^i u^k + p\, g^{ik} + S^{ik} + u^k q^i + u^i q^k, \tag{1.199}$$

where ε is the total energy density, p is the pressure,

$$S_{ik} = \nu \rho \sigma_{ik}, \tag{1.200}$$

is the viscous stress tensor, ρ is the rest mass density, and q^i is the radiative energy flux. In the last equation, ν is the kinematic viscosity coefficient and σ_{ik} is the shear tensor of the velocity field. From the first law of thermodynamic, it follows that

$$d\varepsilon = \frac{\varepsilon + p}{\rho} d\rho + \rho T dS, \tag{1.201}$$

where T is the temperature and S is the entropy per unit mass. Note that in the physical units, $\varepsilon = \rho c^2 + \Pi$, where Π is the internal energy. For nonrelativistic fluids, $\Pi \ll \rho c^2$ and $p \ll \rho c^2$, and therefore

$$\varepsilon + p \approx c^2 \rho. \tag{1.202}$$

We shall use this approximation (in geometrical units $\varepsilon + p \approx \rho$) in all our calculations. This approximation does not automatically ensure that the sound speed is below c, and one should check this a posteriori when models are constructed. We write the first law of thermodynamics in the form:

$$dU = -p\,d\left(\frac{1}{\rho}\right) + T dS, \tag{1.203}$$

where $U = \Pi/\rho$.

1.8 Slim-Disc Equations in Kerr Geometry

General-relativistic effects play an important role in the physics of thin ($H/r \ll 1$) accretion discs close to the black hole, but they determine the properties of slim ($H/r \lesssim 1$) discs. We will derive the slim-disc equation, and before discussing their properties, we will say few words about thin discs.

It is convenient to write the final form of all the slim-disc equations at the equatorial plane, $z = 0$. Only these equations which do not refer to the vertical structure could be derived directly from the quantities at the equatorial plane with no further approximations. All other equations are approximated—either by expansion in terms of the relative disc thickness H/r, or by vertical averaging.

1.8.1 Mass Conservation Equation

From general equation of mass conservation,

$$\nabla^i (\rho u_i) = 0, \tag{1.204}$$

and definition of the surface density Σ,

$$\Sigma = \int_{-H(r)}^{+H(r)} \rho(r, z) dz \approx 2H\rho, \tag{1.205}$$

we derive the mass conservation equation,

$$\dot{M} = -2\pi \Delta^{1/2} \Sigma \frac{V}{\sqrt{1 - V^2}}. \tag{1.206}$$

In the Newtonian limit, the mass conservation equation is as follows:

$$\dot{M} = -2\pi \Sigma v_r. \tag{1.207}$$

1.8.2 Equation of Angular Momentum Conservation

From the general form of the angular momentum conservation,

$$\nabla_k \left(T^{ki} \xi_i \right) = 0, \tag{1.208}$$

we derive, after some algebra,

$$\frac{\dot{M}}{2\pi r} \frac{d\mathfrak{L}}{dr} + \frac{1}{r} \frac{d}{dr} \left(\Sigma \nu A^{3/2} \frac{\Delta^{1/2} \gamma^3}{r^4} \frac{d\Omega}{dr} \right) - F^- \mathfrak{L} = 0, \tag{1.209}$$

where $F^- = 2q_z$ is the vertical flux of radiation and

$$\mathfrak{L} \equiv -(u\xi) = -u_\varphi = \gamma \left(\frac{A^{3/2}}{r^3 \Delta^{1/2}} \right) \tilde{\Omega}, \tag{1.210}$$

is the specific (per unit mass) angular momentum. The term $F^- \mathscr{L}$ represents angular momentum losses through radiation. Although it was always fully recognized that angular momentum may be lost this way, it has been argued that this term must be very small. Rejection of this term enormously simplifies numerical calculations, because with $F^- \mathfrak{L} = 0$ Eq. (1.209) can be trivially integrated,

$$\frac{\dot{M}}{2\pi} (\mathfrak{L} - \mathfrak{L}_0) = -\Sigma \nu A^{3/2} \frac{\Delta^{1/2} \gamma^3}{r^4} \frac{d\Omega}{dr} \equiv \frac{\mathfrak{T}}{2\pi}, \tag{1.211}$$

where $\mathfrak{L}_0$ is the specific angular momentum of matter *at the horizon* ($\Delta = 0$). In the numerical scheme for integrating the slim Kerr equations (with $F^- \mathfrak{L}$ assumed to be zero), the quantity $\mathfrak{L}_0$ plays an important role: it is the eigenvalue of the solutions that passes regularly through the sonic point. The rhs of Eq. (1.211) $\mathfrak{T}$ represent the viscous torque transporting angular momentum.

> In the Newtonian case, a geometrically thin disc is Keplerian $\Omega \approx \Omega_K$ (see Eq. (1.214)), $\mathfrak{L}_K = \ell_K = R^2 \Omega_K$, and Eq. (1.211) takes the familiar form of
>
> $$\nu \Sigma = \frac{\dot{M}}{3\pi} \left[1 - \frac{\ell_0}{\ell_K} \right],$$
>
> (see Eq. 1.57).

1.8.3 Equation of Momentum Conservation

From the r-component of the equation $\nabla_i T^{ik} = 0$, one derives

$$\frac{V}{1-V^2}\frac{dV}{dr} = \frac{\mathscr{A}}{r} - \frac{1}{\Sigma}\frac{dP}{dr}, \tag{1.212}$$

where $P = 2Hp$ is the vertically integrated pressure and

$$\mathscr{A} = -\frac{MA}{r^3 \Delta \Omega_k^+ \Omega_k^-}\frac{(\Omega - \Omega_k^+)(\Omega - \Omega_k^-)}{1 - \tilde{\Omega}^2 \tilde{R}^2}. \tag{1.213}$$

Note that in Eq. (1.212), the viscous term has been neglected.

The Newtonian limit of Eq. (1.212) is

$$v_r \frac{dv_r}{dr} - \left(\Omega^2 - \Omega_K^2\right)r + \frac{c_s^2}{r} = 0 \tag{1.214}$$

For a thin disc: $H/r \approx c_s^2/r\Omega_K \ll 1$, Eq. (1.214), is simply $\Omega \approx \Omega_K$; i.e. a thin Newtonian disc is Keplerian. The thickness of the disc depends on the efficiency of radiative processes: efficient radiative cooling implies a low speed of sound.

1.8.4 Equation of Energy Conservation

From the general form of the energy conservation,

$$\nabla_i \left(T^{ik}\eta_k\right) = 0, \tag{1.215}$$

and the first law of thermodynamics,

$$T = \frac{1}{\rho}\left(\frac{\partial \varepsilon}{\partial S}\right)_\rho, \quad p = \rho \left(\frac{\partial \varepsilon}{\partial \rho}\right)_S - \varepsilon, \tag{1.216}$$

the energy equation can be written in general as

$$Q^{\text{adv}} = Q^+ - Q^-, \tag{1.217}$$

where

$$Q^+ = \nu \Sigma \frac{A^2}{r^6}\gamma^4 \left(\frac{d\Omega}{dr}\right)^2 \tag{1.218}$$

is the surface viscous heat generation rate, Q^- is the radiative cooling flux (both surfaces) which is discussed in Sect. 1.5, and Q^{adv} is the advective cooling rate due to the radial motion of the gas. It is expressed as

$$Q^{\mathrm{adv}} = \frac{\Sigma V}{\sqrt{1 - V^2}} \frac{\Delta^{1/2}}{r} T \frac{dS}{dr} \equiv -\frac{\dot{M}}{2\pi r} T \frac{dS}{dr}. \tag{1.219}$$

In *stationary* accretion flows, advection is important only in the inner regions close to the compact accretor. In the rest of the flow, the energy equation is just

$$Q^+ = Q^-. \tag{1.220}$$

In the newtonian limit of Eqs. (1.218) and (1.211), one obtains

$$Q^+ = \frac{3}{8\pi} \frac{GM\dot{M}}{R^3} \left(1 - \frac{\ell_0}{\ell}\right). \tag{1.58}$$

1.8.5 Equation of Vertical Balance of Forces

The equation of vertical balance is obtained by projecting the conservation equation onto the θ direction

$$h_\theta^i \nabla_k T_i^k = 0, \qquad \text{where} \qquad h_\theta^i = \delta_\theta^i - u^i u_\theta \tag{1.221}$$

and neglecting the terms $\mathcal{O}^3(\cos\theta)$. For a nonrelativistic fluid, this leads to

$$\frac{dP}{dz} = -\rho g_z z = -\rho \frac{\mathcal{L}^2 - a^2 \left(\mathfrak{E}^2 - 1\right)}{r^4} z. \tag{1.222}$$

In the newtonian limit, Eq. (1.222) becomes

$$\frac{dP}{dz} = -\rho \frac{\ell_K^2}{r^4} z \qquad \text{(see Eq. 1.18)}.$$

1.9 The Sonic Point and the Boundary Conditions

1.9.1 The "No-Torque Condition"

There have been a lot of discussion about the inner boundary condition in an accretion disc. The usual reasoning is that for a thin disc, the inner boundary is at ISCO and since it is where circular orbits end the boundary condition should be simply that the "viscous" torque vanishes (there is no orbit below the ISCO to interact with). Several authors have challenged this conclusion, but a very simple argument by Bohdan Paczyński [44] shows the fallacy of these challenges.

Using Eq. (1.206), one obtains from Eq. (1.211)

$$v^{(r)} = v \frac{A^{3/2}\gamma^2}{r^4} \frac{1}{\mathfrak{L} - \mathfrak{L}_0} \cdot \frac{d\Omega}{dr} \tag{1.223}$$

Next, from the viscosity prescription $v \approx \alpha H^2 \Omega$, and taking for simplicity the nonrelativistic approximation (this does not affect the validity of the argument but allows skipping irrelevant in this context multiplicative factors), one can write

$$v_r \approx \alpha \, H^2 \, \frac{\ell}{\ell - \ell_0} \frac{d\Omega}{dr} \approx \alpha \, H^2 \, \frac{\ell}{\ell - \ell_0} \, \frac{\Omega}{r} \approx \alpha \, v_\varphi \left(\frac{H}{R}\right)^2 \frac{\ell}{\ell - \ell_0}, \tag{1.224}$$

where $v_\varphi = R\Omega$. Although we have dropped the GR terms, the Eq. (1.224) does not assume that the radial velocity is small; i.e. this equation holds within the disc as well as within the stream below the ISCO.

Far out in the disc, where $\ell \gg \ell_0$, one obtains the standard formula (see Eq. 1.36)

$$v_r \approx \alpha \, v_\varphi \left(\frac{H}{R}\right)^2, \qquad R \gg R_{in}. \tag{1.225}$$

The flow crosses the black hole surface at the speed of light, and since it is subsonic in the disc, it must somewhere become transonic, i.e. to go through a sonic point, close to disc's inner edge.

At the sonic point, we have $v_r = c_s \approx (H/R)v_\varphi$, and the Eq. (1.224) becomes:

$$\frac{v_r}{c_s} = 1 \approx \alpha \, \frac{H_{in}}{R_{in}} \, \frac{\ell_{in}}{\ell_{in} - \ell_0}, \qquad R = R_{in} \tag{1.226}$$

If the disc is thin, i.e. $H_{in}/R_{in} \ll 1$, and the viscosity is small, i.e. $\alpha \ll 1$, then Eq. (1.226) implies that $(\ell_{in} - \ell_0)/\ell_{in} \ll 1$; i.e. the specific angular momentum at the sonic point is almost equal to the asymptotic angular momentum at the horizon.

In a steady-state disc, the torque $\mathfrak{T}$ has to satisfy the equation of angular momentum conservation (1.211), which can be written as

$$\mathfrak{T} = \dot{M} \, (\ell - \ell_0), \qquad \mathfrak{T}_{in} = \dot{M} \, (\ell_{in} - \ell_0). \tag{1.227}$$

Thus, it is clear that for a thin, low viscosity disc, the 'no-torque inner boundary condition' ($\mathfrak{T}_{in} \approx 0$) is an excellent approximation *following from angular momentum conservation*.

However, if the disc and the stream are thick, i.e. $H/r \sim 1$, and the viscosity is high, i.e. $\alpha \sim 1$, then the angular momentum varies also in the stream in accordance with the simple reasoning presented above. However, the no-stress condition at the disc inner edge might be not satisfied.

Additional reading: Reference [6].

Acknowledgments I am grateful to Cosimo Bambi for having invited me to teach at the 2014 Fudan Winter School in Shanghai. Discussions with and advice of Marek Abramowicz, Tal Alexander, Omer Blaes and Olek Sądowski were of great help. I thank the Nella and Leon Benoziyo Center for Astrophysics at the Weizmann Institute for its hospitality in December 2014/January 2015 when parts of these lectures were written. This work has been supported in part by the French Space Agency CNES and by the Polish NCN grants DEC-2012/04/A/ST9/00083, UMO-2013/08/A/ST9/00795 and UMO-2015/19/B/ST9/01099.

Appendix

Thermodynamical Relations

The equation of state can be expressed in the form:

$$P = P_r + \frac{\mathscr{R}}{\mu_i}\rho T_i + \frac{\mathscr{R}}{\mu_e}\rho T_e + \frac{B^2}{24\pi}, \qquad (1.228)$$

where P_r is the radiation pressure, $\mathscr{R}$ is the gas constant, μ_i and μ_e are the mean molecular weights of ions and electrons respectively, T_i, and T_e are ion and electron temperatures, a is the radiation constant (not to be confused with the dimensionless angular momentum a in the Kerr metric), and B is the intensity of a isotropically tangled magnetic field, includes the radiation, gas and magnetic pressures. The radiation pressure P_r, the gas pressure P_g and the magnetic pressure P_m correspond respectively to the first term, the second and third terms, and the last term in Eq. (1.228).

The mean molecular weights of ions and electrons can be well approximated by:

$$\mu_i \approx \frac{4}{4X + Y}, \qquad \mu_e \approx \frac{2}{1 + X}, \qquad (1.229)$$

where X is the relative mass abundance of hydrogen and Y that of helium. We may define a temperature as

$$T = \mu \left(\frac{T_i}{\mu_i} + \frac{T_e}{\mu_e} \right), \qquad (1.230)$$

where

$$\mu = \left(\frac{1}{\mu_i} + \frac{1}{\mu_e} \right)^{-1} \approx \frac{2}{1 + 3X + 1/2Y} \qquad (1.231)$$

is the mean molecular weight. In the case of a one-temperature gas ($T_i = T_e$), one has $T = T_i = T_e$. For an optically thick gas, $P_r = (4\sigma/3c)T_r^4$.

For the frozen-in magnetic field pressure $P_m \sim B^2 \sim \rho^{4/3}$, therefore, we may write the internal energy as

$$U = \frac{4\sigma}{\rho c}T_r^4 + \frac{\mathscr{R}T}{\mu m_u(\gamma_g - 1)} + e_o\rho^{1/3}, \tag{1.232}$$

where e_o is a constant ($P_m = 1/3e_o\rho^{4/3}$) and γ_g is the ratio of the specific heats of the gas. We define

$$\beta = \frac{P_g}{p}, \quad \beta_m = \frac{P_g}{P_g + P_m}, \quad \beta^* = \frac{4 - \beta_m}{3\beta_m}\beta. \tag{1.233}$$

From Eqs. (1.228) and (1.232), one obtains the following formulae (see, e.g. "Cox and Giuli" 2004) for the specific heat at constant volume:

$$c_V = \frac{\mathscr{R}}{\mu(\gamma_g - 1)}\left[\frac{12(1 - \beta/\beta_m)(\gamma_g - 1) + \beta}{\beta}\right] = \frac{4 - 3\beta^*}{\Gamma_3 - 1}\frac{P}{\rho T} \tag{1.234}$$

and the adiabatic indices:

$$\Gamma_3 - 1 = \frac{(4 - 3\beta^*)(\gamma_g - 1)}{12(1 - \beta/\beta_m)(\gamma_g - 1) + \beta} \tag{1.235}$$

$$\Gamma_1 = \beta^* + (4 - 3\beta^*)(\Gamma_3 - 1). \tag{1.236}$$

The ratio of specific heats is $\gamma = c_p/c_V = \Gamma_1/\beta$. For $\beta = \beta_m$, we have $\Gamma_3 = \gamma_g$ and $\Gamma_1 = (4 - \beta)/3 + \beta(\gamma_g - 1)$. For an equipartition magnetic field ($\beta = 0.5$), one gets $\Gamma_1 = 1.5$ and for $\beta = 0.95$, $\Gamma_1 = 1.65$ (here, we have used $\gamma_g = 5/3$). One expects $\beta_m \sim 0.5 - 1$. Since

$$T\frac{dS}{dR} = c_V\left[\frac{d \ln T}{dR} - (\Gamma_3 - 1)\left(\frac{d \ln \Sigma}{dR} - \frac{d \ln H}{dR}\right)\right], \tag{1.237}$$

the advective flux is written in the form:

$$Q^{\text{adv}} = \frac{\dot{M}}{2\pi R^2}\frac{P}{\rho}\xi_a \tag{1.238}$$

where

$$\xi_a = -\left[\frac{4 - 3\beta^*}{\Gamma_3 - 1}\frac{d \ln T}{d \ln R} + (4 - 3\beta^*)\frac{d \ln \Sigma}{d \ln R}\right]. \tag{1.239}$$

The term $\propto d \ln H/d \ln R$ has been neglected. Since no rigorous vertical averaging procedure exists, the presence or not of the $d \ln H/d \ln R$—type terms in this (and other) equation may be decided only by comparison with 2D calculations.

The formulae derived in this section are valid for the optically thin case $\tau = 0$ if one assumes $\beta = \beta_m$.

REFERENCE: Weiss, A., Hillebrandt, W., Thomas, H.-C., and Ritter, H. 2004, *Cox and Giuli's Principles of Stellar Structure*, Cambridge, UK: Princeton Publishing Associates Ltd, 2004.

References

1. M.A. Abramowicz, *Growing Black Holes: Accretion in a Cosmological Context*, ESO Astrophysics Symposia (Berlin, Springer, 2005)
2. M.A. Abramowicz, W. Kluźniak, Ap&SS **300**, 127 (2005)
3. M.A. Abramowicz, B. Czerny, J.P. Lasota, E. Szuszkiewicz, ApJ **332**, 646 (1988)
4. M.A. Abramowicz, X. Chen, S. Kato, J.-P. Lasota, O. Regev, ApJ **438**, L37 (1995)
5. M.A. Abramowicz, X.-M. Chen, M. Granath, J.-P. Lasota, ApJ **471**, 762 (1996)
6. N. Afshordi, B. Paczyński, ApJ **592**, 354 (2003)
7. S.A. Balbus, in *Physical Processes in Circumstellar Disks around Young Stars*, ed. by P.J.V. Garcia (University of Chicago Press, Chicago, 2011), p. 237. arXiv:0906.0854
8. S.A. Balbus, J.F. Hawley, ApJ **376**, 214 (1991)
9. O. Blaes, Space Sci. Rev. **183**, 21 (2014)
10. A.E. Broderick, T. Johannsen, A. Loeb, D. Psaltis, ApJ **784**, 7 (2014)
11. M.S.B. Coleman, I. Kotko, O. Blaes, J.-P. Lasota, MNRAS, submitted (2015)
12. S. Collin, J.-P. Zahn, A&A **477**, 419 (2008)
13. M. Coriat, R.P. Fender, G. Dubus, MNRAS **424**, 1991 (2012)
14. J.P. Cox, R.T. Giuli, *Principles of Stellar Structure* (Gordon & Breach, New York, 1968)
15. G. Dubus, J.-P. Lasota, J.-M. Hameury, P. Charles, MNRAS **303**, 139 (1999)
16. G. Dubus, J.-M. Hameury, J.-P. Lasota, A&A **373**, 251 (2001)
17. A.A. Esin, J.-P. Lasota, R.I. Hynes, A&A **354**, 987 (2000)
18. J. Frank, A. King, D.J. Raine, *Accretion Power in Astrophysics* (Cambridge University Press, Cambridge, 2002)
19. C.F. Gammie, ApJ **553**, 174 (2001)
20. J. Goodman, MNRAS **339**, 937 (2003)
21. J.-M. Hameury, K. Menou, G. Dubus, J.-P. Lasota, J.-M. Hure, MNRAS **298**, 1048 (1998)
22. S. Hirose, O. Blaes, J.H. Krolik, M.S.B. Coleman, T. Sano, ApJ **787**, 1 (2014)
23. K. Horne, M.C. Cook, MNRAS **214**, 307 (1985)
24. Y.-F. Jiang, J.M. Stone, S.W. Davis, ApJ **778**, 65 (2014)
25. Y.-F. Jiang, J.M. Stone, S.W. Davis, ApJ **796**, 106 (2014)
26. S. Kato, J. Fukue, S. Mineshige, *Black-Hole Accretion Disks – Towards a New Paradigm* (Kyoto University Press, Kyoto, 2008)
27. A.R. King, MNRAS **393**, L41 (2009)
28. A. King, C. Nixon, Class. Quantum Gravity **30**, 244006 (2013)
29. I. Kotko, J.-P. Lasota, A&A **545**, 115 (2012)
30. J.-P. Lasota, in *Theory of Accretion Disks - 2*, vol. 417, NATO Advanced Science Institutes (ASI) Series C, ed. by W.J. Duschl (Kluwer, Dordrecht, 1994), p. 341
31. J.-P. Lasota, New Astron. Rev. **45**, 449 (2001)
32. J.-P. Lasota, G. Dubus, K. Kruk, A&A **486**, 523 (2008)

33. J.-P. Lasota, A.R. King, G. Dubus, ApJL **801**, L4 (2015)
34. L.D. Landau, E.M. Lifshitz, *Fluid Mechanics; Course of Theoretical Physics* (Pergamon Press, Oxford, 1987)
35. D.N.C. Lin, J.E. Pringle, MNRAS **225**, 607 (1987)
36. K. Menou, J.-M. Hameury, R. Stehle, MNRAS **305**, 79 (1999)
37. F. Meyer, E. Meyer-Hofmeister, A&A **104**, L10 (1981)
38. R. Narayan, I. Yi, ApJ **428**, L13 (1994)
39. I.D. Novikov, K.S. Thorne, in *Black holes (Les astres occlus)*, École de Houches, ed. by C. DeWitt, B.S. DeWitt (Gordon & Breach, 1972), p. 343
40. G.I. Ogilvie, MNRAS **304**, 557 (1999)
41. B. Paczyński, AcA **19**, 1 (1969)
42. B. Paczyński, ApJ **216**, 822 (1977)
43. B. Paczyński, AcA **28**, 91 (1978)
44. B. Paczyński (2000). arXiv:astro-ph/0004129
45. J.C.B. Papaloizou, J.E. Pringle, MNRAS **202**, 118 (1983)
46. Partnership ALMA, C.L. Brogan, L.M. Perez et al. ApJL, **808**, L3 (2015). arXiv:1503.02649
47. J. Poutanen, G. Lipunova, S. Fabrika, A.G. Butkevich, P. Abolmasov, MNRAS **377**, 1187 (2007)
48. D. Prialnik, *An Introduction to the Theory of Stellar Structure and Evolution* (Cambridge University Press, Cambridge, 2009)
49. A. Sądowski, ApJS **183**, 171 (2009)
50. A. Sądowski, PhD Thesis (CAMK) (2011) arXiv:1108.0396
51. A. Sądowski, R. Narayan, MNRS **453**, 3213 (2015). arXiv:1503.00654
52. A. Sądowski, R. Narayan, J.C. McKinney, A. Tchekhovskoy, MNRAS **439**, 503 (2014)
53. N.I. Shakura, R.A. Sunyaev, A&A **24**, 337 (1973)
54. J. Smak, AcA **49**, 391 (1999)
55. H.C. Spruit (2010). arXiv:1005.5279
56. A. Toomre, ApJ **139**, 1217 (1964)
57. F.H. Vincent, T. Paumard, G. Perrin et al., The Galactic Center: a Window to the Nuclear Environment of Disk Galaxies **439**, 275 (2011)
58. J. Wood, K. Horne, G. Berriman et al., MNRAS **219**, 629 (1986)
59. Z. Yan, W. Yu (2014), ApJ, **805**, 87 (2015) arXiv:1408.5146
60. F. Yuan, R. Narayan, ARA& A **52**, 529 (2014)

Chapter 2
Transient Black Hole Binaries

Tomaso M. Belloni and Sara E. Motta

Abstract The last two decades have seen a great improvement in our understanding of the complex phenomenology observed in transient black hole binary systems, especially thanks to the activity of the Rossi X-ray Timing Explorer satellite, complemented by observations from many other X-ray observatories and ground-based radio, optical and infrared facilities. Accretion alone cannot describe accurately the intricate behaviour associated with black hole transients, and it is now clear that the role played by different kinds of (often massive) outflows seen at different phases of the outburst evolution of these systems is as fundamental as the one played by the accretion process itself. The spectral-timing states originally identified in the X-rays and fundamentally based on the observed effect of accretion have acquired new importance as they now allow to describe within a coherent picture the phenomenology observed at other wavelength, where the effects of ejection processes are most evident. With a particular focus on the phenomenology seen in the X-ray band, we review the current state of the art of our knowledge of black hole transients, describing the accretion–ejection connection at play during outbursts through the evolution of the observed spectral-timing properties. Although we mainly concentrate on the observational aspects of the global X-ray transient picture, we also provide physical insight to it by reviewing (when available) the theoretical explanations and models proposed to explain the observed phenomenology.

2.1 Introduction

Black hole (BH) binaries (hereafter BHBs) are binary systems consisting of a non-collapsed star and a black hole. They provide a unique laboratory for the study not only of accretion of matter onto a compact object, but also of the strongly

T.M. Belloni (✉)
INAF - Osservatorio Astronomico di Brera, via E. Bianchi 46, I-23807 Merate, Italy
e-mail: tomaso.belloni@brera.inaf.it

S.E. Motta
Department of Physics, Astrophysics, University of Oxford,
Denys Wilkinson Building, Keble Road, Oxford OX1 3RH, UK
e-mail: sara.motta@physics.ox.ac.uk

© Springer-Verlag Berlin Heidelberg 2016
C. Bambi (ed.), *Astrophysics of Black Holes*, Astrophysics
and Space Science Library 440, DOI 10.1007/978-3-662-52859-4_2

curved spacetime in the vicinity of a black hole, where the effects of general relativity in a strong gravitational field cannot be ignored and can be studied (see [137]).

The first BHB, Cygnus X-1, was discovered in the early years of X-ray astronomy with instruments on board sounding rockets. It became the first candidate for systems hosting a black hole when optical observations revealed its binary nature and allowed the estimate of the mass of the compact object (see Chap. 3 in this book). Its X-ray properties included strong variability on short timescales and a very hard energy spectrum. Cyg X-1 is a persistent system, which we know now are rather rare. Most BHBs are transient, difficult to discover in the absence of a wide-field X-ray instrument. The first such system, A 0620-00, was discovered in 1975 with the Ariel V satellite [47] and in 1986. It became a much stronger candidate for hosting a black hole, with a minimum mass of 3.2 $M_\odot$ for the compact object [93]. A few additional persistent systems were identified, notably the two in the LMC and the galactic source GX 339-4, but it was only with the Japanese satellite Ginga that new transients were discovered and followed, thanks to the availability of an all-sky monitor and of a large-area detector. Systems such as GX 339-4 and Cyg X-1 were known to undergo transitions in their flux, spectral and timing properties, which had led to the definition of a high (flux) and low state (see, e.g. [169]). The detailed study of several observations, in particular of the transient GS 1124-683 (also known as Nova Muscae 1991), GX 339-4 and Cyg X-1 led to the identification of complex spectral/timing properties that in turn led to the definition of additional states (see [17, 19, 107, 110]). In particular, the first spectral studies with accretion disc models opened the way to techniques to measure the inner radius of the disc. In the timing domain, quasi-periodic oscillations (QPOs) were discovered. These features yield discrete frequencies that constitute a very direct measurement of timescales in the system, which are associated with the accretion phenomenon and/or to effects of general relativity. A full review of the observational status after Ginga can be found in [168]), where essentially all known information on BHBs at the time is recorded.

The situation changed dramatically starting from the second half of the nineties. At the end of 1995, the Rossi X-ray Timing Explorer (RXTE) was launched and opened a new window onto phenomena of fast variability from neutron star and black hole binaries. Its All-Sky Monitor (ASM) allowed the timely discovery of outbursts of old and new transients and the large-area proportional counter array (PCA), operated in a very flexible way, accumulated high signal data for timing analysis and spectral analysis at moderate spectral resolution. A few years later, the availability of high-spectral-resolution instruments on board Chandra and XMM-Newton opened the way to perform detailed spectral studies, in particular on relativistically skewed emission lines. Overall, there was a real explosion of information and now, after the demise of RXTE and with the availability of further missions such as INTEGRAL, Swift, Suzaku and NuSTAR, it is possible to combine archival data and new observations to obtain new insight on the emission properties of BHBs. This allows us to understand much better the process of accretion onto a black hole and new exciting results are possible that were not even imaginable only a few years ago. The long-term evolution of the X-ray emission, in particular for transients, is now clear in its phenomenology and provides a solid framework upon which to study detailed spectral and variability properties. Direct effects of general relativity are studied and measured through the analysis of broadband spectra, emission lines and QPOs (see below and Chap. 3), the properties of continuum emission are interpreted with physical models and global models for different states are being proposed.

At the same time, the discovery of relativistic jets from BHBs [105, 106] (see also [50, 55]) opened a new path to the study of accretion. The difference in timescale and count statistics with active galactic nuclei, where jets were known since much earlier, allows the study of the connection between the accretion onto the compact object and ejection of relativistic jets, introducing a new window to understand the physics of accretion and to unveil the mechanism responsible for the ejections (see [50, 55]). Moreover, in the past few years the existence of powerful non-relativistic winds has been discovered in a number of objects (see [50]). These winds appear to be flowing in a direction parallel to the accretion disc, as opposed to jets, and to be mutually exclusive with the jets. The full picture of an accreting black hole is now much different from that of only a couple of decades ago, when observations in the X-ray band alone and only at low-energy resolution made us miss very important components (jets and winds) which can be energetically very important. Jets and outflows are the subject of Sect. 2.3.

In addition to the comparison with active galactic nuclei, the supermassive counterparts of stellar mass accreting black holes, the strong similarities between the BHBs and other systems at subgalactic scale are being studied and exploited. Neutron star binaries share many properties in the X-ray band and, when their magnetic field is sufficiently low, also in the radio, where relativistic jets have been observed [102, 119]. Binaries in external galaxies are also accessible with current instrumentation, although naturally at lower count rates, and in particular the enigmatic ultraluminous X-ray sources (ULXs) are being studied in order to understand whether they contain intermediate-mass black holes or are super-Eddington counterparts of our galactic objects (see [53]). In this respect, a detailed comparison with BHBs is instrumental to unveil this mystery.

In this review, we aim at presenting the current state of the art of BHB research, concentrating on the properties of accretion and ejection, while the Chap. 3 addresses the issue of measurement of physical quantities of the central object. The sheer amount of information available today makes it impossible to cover all aspects. Many specific and additional details can also be found, e.g., in [24, 49, 50, 85].

2.2 X-ray Emission

In BHBs, the X-ray emission originates from the inner regions of the accretion flow and, possibly, from the base of the relativistic jets [90]. Even restricting ourselves to the two original source states, low/hard and high/soft, the details of the emission appear to be very complex (see, e.g. [62]). Both the energy spectra and the fast variability contain multiple components, which vary in a correlated fashion. While the details of these properties are given in the next sections, it is important to understand the regularities that exist in the time evolution of the observables. Before the mid-1990s, only a few sources were available and the coverage was not sufficient to study these aspects in sufficient detail; only with RXTE, it has been possible to find a general scheme for the characterization of source states. Even before then, it was clear that a simple classification of the observed properties on the basis of energy spectra alone was not possible and fast variability had to be included in the picture. In this sense, the notion of "spectral states" does not carry any meaning.

While a few BHBs are persistent sources, like the archetypal system Cyg X-1 (always accreting at a high rate and emitting luminosities above 10^{37} erg/s) most of them are of transient nature. They spend most of the time in a low accretion regime ($L_X < 10^{33}$ erg/s), where observations are still limited by the low number of counts (see [132] and references therein). With a recurrence period that varies between several months and decades, the accretion rate onto the central objects increases by orders of magnitude and the sources go into outburst for a time that can range from a few days to, more commonly, several months (one peculiar object, GRS 1915+105 is at present active since 23 years). Their X-ray luminosity increases, peaks and then decreases and can roughly be adopted as a proxy for accretion rate, while the detailed properties of the energy spectra and fast variability change, at times in a very abrupt way. It was only at the beginning of the last decade, however, that a coherent picture emerged, which can be applied to most systems [65, 69].

Outbursts of different systems and even multiple outbursts from the same object have time evolutions which can differ considerably (see Belloni [20]). However, when the evolution of an outburst is represented in a hardness–intensity diagram (HID), strong regularities emerge. A HID is equivalent to a hardness–magnitude diagram in the optical band: on the abscissa is the ratio of counts in two separate bands (hard/soft), which gives a rough indication of the hardness of the energy spectrum, and on the ordinate the total count rate over a broad energy band, a proxy for luminosity and accretion rate [20, 24, 66, 69]. As such, the diagram is source dependent (because of interstellar absorption) and instrument dependent, but it is extremely useful to follow BHB outbursts. An extension to an independent form of the diagram (with flux ratio between the components and total flux respectively) has been proposed [46]: this has the advantage of containing physical quantities, but the disadvantage of being insensitive to small changes when one component dominates, besides being of course model-dependent. An example of a HID based on RXTE/PCA data for the best known system, GX 339-4, is shown in the top left panel of Fig. 2.1. Here, four outbursts are plotted (2002, 2004, 2007, 2010). The general evolution is the same: a "q"-shaped diagram travelled counterclockwise from the bottom right (faint and hard). Quiescence cannot be included as the RXTE/PCA instrument was not sensitive to low-flux observations, but we know that the hard branch softens as flux decreases (see [7] and references therein). From the HID, one can easily identify the two historical states as the two "vertical" branches. The hard branch extending to the stem of the "q" corresponds to the low/hard state (LHS), which is observed at the start and at the end of an outburst only, never in the middle. The branch is not really vertical as the logarithmic axis suggests, but there is a marked softening as the source brightens (see, e.g. [112]). The left branch corresponds to the high/soft state (HSS). The scatter of the points is magnified by the log scale and contains some excursions back to other states (see below) as well as intrinsic variability of the hard spectral component. The transition between these two states takes place at two different flux levels. At high flux, the source moves from LHS to HSS and at low flux it returns to the LHS, completing a hysteresis cycle (originally identified by [109], see also [85, 88]). Two additional diagrams have been proven useful for following the evolution of an outburst and identify source states. The first is the HRD (hardness–rms diagram, bottom left panel in Fig. 2.1), where the Y-axis contains the fractional rms variability integrated over a broad range of frequencies (see [20]). Remarkably, no hysteresis is observed in this diagram: at each hardness corresponds a single value of fractional rms, regardless of the flux. The second is the RID (rms–intensity diagram, right panel in Fig. 2.1), the third combination of the same three observables. Here the hysteresis, as expected, is clearly present and the location of specific states is rather precise (see [120]). The central part of the diagram identifies two additional states, the hard intermediate state (HIMS) and soft intermediate state (SIMS). The transition between the states corresponds to precise values in hardness, obviously source and instrument dependent. The identification of these thresholds is based on the properties of fast variability and/or changes in the multi-wavelength relations. For a precise determination of the different states, we refer the reader to [24, 120], here we include a brief summary:

- Low/Hard State (LHS). The LHS has only been observed in the first and last stages of an outburst. In some cases, the first RXTE/PCA observations found a source already in a softer state, but the initial LHS might have been too fast to observe. It corresponds to the right branch in the HID and to the straight diagonal line in the RID (see Fig. 2.1). It is characterized by large variability (around 40 % fractional rms in the case of GX 339-4) and a hard spectrum (see below). In transients, it is only observed at the start and end of outbursts (although at times the start LHS is missed altogether) and is the most common state for the persistent system Cyg X-1. The variability is in the form of broadband noise made of a few components whose characteristic frequencies increase with luminosity (and decreasing hardness).
- Hard Intermediate State (HIMS). In the HID, this state corresponds to a large part of the area between the LHS and the HSS, covering the horizontal tracks, both at high and low flux. This state appears after the initial LHS and reappears before the source goes back to the LHS at the end of the outburst. In addition, secondary transitions to and from it can be observed (see below). The softening compared to the LHS is due to two effects: the appearance in the observational range of flux from the thermal disc and the steepening of the hard component (see below and [112]). The fast variability is an extension of that of the LHS, with characteristic frequencies increasing and total fractional rms decreasing (see the HRD and the RID in Fig. 2.1). Type-C QPOs are present (see below).
- Soft Intermediate State (SIMS). The energy spectrum is slightly softer than the HIMS, putting this state to the left of the HIMS in the HID. In the HID and RID

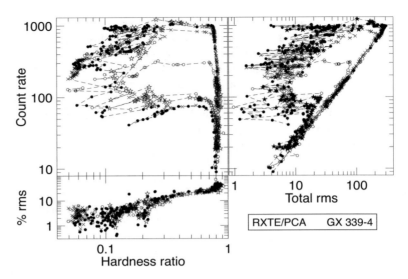

Fig. 2.1 The three main diagrams for the representation of the X-ray evolution of black hole binaries for three outbursts of GX 339-4 as observed with RossiXTE. *Top left* Hardness–intensity diagram (HID), *top right* rms–intensity diagram (RID) and *bottom left* hardness–rms diagram (HRD)

these points are not immediately identifiable, but can be seen in the HRD (bottom panel in Fig. 2.1) as a cloud of points at a lower rms than the main branch (around hardness 0.2). While the energy spectrum below 10 keV is very similar to that of the softer HIMS points, at high energies the spectrum unlike the HIMS does not show a significant high-energy cut-off [112]. The identifying feature of this state is the disappearance of the band-limited noise components in the power density spectrum, replaced by a weaker power law component (hence, the lower fractional rms) and the appearance of a marked type-B QPO.

- High–Soft State (HSS). The spectrum is soft, dominated by an optically thick accretion disc, variability is low. Occasional low-frequency QPOs can be detected, identified with type-C (see [113]). This state is reached from the intermediate states and left through the intermediate states.

Although for different sources diagrams can look different in the HID in Fig. 2.1, the sequence of states from quiescence to quiescence is the following: LHS—HIMS—SIMS—HSS—(minor transitions to and from HIMS and SIMS)—HSS—SIMS—HIMS—LHS. Some transient remained in the LHS throughout the outburst, a few showed failed transitions in the sense that the LHS–HIMS sequence was not followed by a transition to the softer states (SIMS and HSS). The few bright persistent sources show a reduced number of states. Cyg X-1 is found mostly in the LHS with transition to the HIMS and possibly all the way to the HSS. LMC X-1 is always in the HSS. LMC X-3 is mostly in the HSS, with brief transitions to the LHS (no strong evidence of intermediate states).

The three diagrams presented above and the state classification are a firm basis upon which to base detailed spectral and timing analysis. One important transition is the HIMS–SIMS one, which appears to be associated to the ejection of relativistic ballistic jets (the crossing of the "jet line", see below).

2.2.1   Energy Spectra

2.2.1.1   The Truncated Disc Model

Long-term X-ray light curves, X-ray spectra, the rapid X-ray variability and the radio jet behaviour have been shown to be consistent with the so-called *truncated disc model*. According to this model, at low luminosities a Shakura–Sunyaev optically thick, geometrically thin accretion disc is truncated at a certain (variable) radius and coexists with a hot, optically thin, geometrically thick accretion flow, which replaces the region between the inner edge of the disc and the innermost stable orbit. Neutron stars are also consistent with the same description [119], but with an additional component due to their surface, giving implicit evidence for the event horizon in black holes.

At low luminosities, the optically thick, geometrically thin disc is truncated at very large radii, being replaced (probably through evaporation, see e.g. [99]) in the inner regions by a hot, inner flow which might also act as the launching site of the jet. Only a few photons from the disc illuminate the flow at this stage; therefore, Compton cooling of the electrons is rather inefficient compared to the heating coming from collisions with protons. The ratio of power in the electrons to that in the seed photons illuminating them—L_h/L_s—is the major parameter (together with the optical depth of the plasma) which determines the shape of a thermal Comptonization spectrum (e.g. [64]). Physically, L_h/L_s sets the energy balance between heating and cooling and, hence, the electron temperature. In the hard state, the relative lack of seed photons illuminating the hot inner flow produces hard thermal Comptonized spectra (with $L_h/L_s >> 1$), roughly characterized by a power law in the 5–20 keV band with photon index $1.5 < \Gamma < 2$ (where the photon spectrum $N(E) \propto E^{-\Gamma}$). Hard spectra of this kind are typical of the LHS.

As the disc moves progressively inwards, it increasingly extends underneath the hot inner flow so that there are more seed photons intercepted by the flow, decreasing L_h/L_s. Therefore, the decrease in disc truncation radius leads to softer spectra, as well as higher frequencies in the power spectra and a faster jet. This results in spectra that are a combination of the hard spectral component described above and the soft spectral component typical of the soft state (see below), i.e. a standard geometrically thin, optically thick accretion disc with a progressively smaller inner radius. These spectra are observed during the (usually) short-lived HIMS and SIMS.

When the truncation radius reaches the innermost stable orbit, the hot flow is thought to collapse into a Shakura–Sunyaev disc and dramatic changes in both the spectral and time domains are seen. This includes a significant decrease in radio flux, as well as the major hard-to-soft spectral transition seen in BHs. The dramatic increase in disc flux due to the presence of the inner disc marks the hard–soft-state transition [48] and also means that any remaining electrons which gain energy outside of the optically thick disc material are subject to much stronger Compton cooling (L_h/L_s is now ≤ 1). This results in much softer Comptonized spectra. Thus, the soft state is characterized by a strong disc and soft tail, roughly described by a power law index of photon index $\Gamma \geq 2$, extending out beyond 500 keV [61]. This tail, differently from the hard tail observed in the LHS, is not produced by thermal Comptonization. In order to extend to 500 keV and beyond, the spectrum should be produced in a region with rather small optical depth and high temperature. However, these conditions would result in a bumpy spectrum, with individual Compton scattering orders separated, in contrast with the observed smooth power law-like tail. Such a spectrum can be instead produced by Compton scattering on a non-thermal electron population, where the index is set predominantly by the shape of the electron distribution rather than L_h/L_s. Soft spectra such as the ones described here are typically seen during the HSS.

As noted in [44], even though there are no observations which unambiguously conflict with the truncated disc models, there exist a tremendous amount of data which can be fit within this geometry (which includes, besides energy spectra, rapid variability characteristics and jet properties). While the truncated disc model is indeed currently a very simplified version of what must be a more complex reality, nonetheless the range of data it can qualitatively explain gives confidence that it captures the essence of the main spectral states.

2.2.1.2 Alternative Geometries

Alternative geometries that include an untruncated disc and (mostly) isotropic source emission have significant problems in matching the observed features of the hard-state spectra.

- The shape of the spectral continuum observed in the hard-state rules out slab corona models as the spectra all peak at high energies. These spectra are possible only if the luminosity in seed photons within the X-ray region is less than that in the hot electrons (i.e. $L_h/L_s \gg 5$), while in a slab geometry only quite small L_h/L_s can be obtained (a disc extending underneath an isotropically radiating hot electron region would intercept around half the Comptonized emission in a slab corona geometry, see [64]).
- A patchy corona allows part of the reprocessed flux to escape without reilluminating the hot electron region, so it can produce the required hard spectra. A patchy corona also allows the reflected flux to escape along with the reprocessed flux, resulting in a strong reflection spectrum for very hard spectra, in direct conflict with the observations (see e.g. [89]).
- Models where the X-rays are produced directly in the jet were proposed by [91]. These produce the hard X-rays by synchrotron emission from the high-energy ex-

tension of the same non-thermal electron distribution which gives rise to the radio emission. However, the observed shape of the high-energy cut-off in the hard state is very sharp and cannot be easily reproduced by synchrotron models [183]. However, there are now composite models where the X-rays are from Comptonization by thermal electrons at the base of the jet (which resides in a hot flow at the centre of a truncated disc), while the radio is from non-thermal electrons accelerated up the jet [92]. This model practically converges onto the truncated disc model, though with some additional weak beaming of the hard X-rays.

- One last alternative to the truncated disc model is represented by the magnetized accretion–ejection model of [54], which has a disc-inner jet structure similar to that by [92], though here the inner, optically thick disc is still present down to the last stable orbit, but with properties very different from the standard Shakura–Sunyaev disc. The transition radius between this jet-dominated disc and the standard accretion disc is variable, producing the range of behaviour seen in the hard state spectra in a similar way to the truncated disc model.

From what has been said above, all currently viable models for the hard state converge on a geometry where the standard disc extends down only to some radius larger than the last stable orbit, with the properties of the flow abruptly changing at this point.

2.2.2 Fast Time Variability

Fast time variability is an important characteristic of BHBs and a key ingredient for understanding the physical processes in these systems. Fast (aperiodic and quasi-periodic) variability is generally studied through the inspection of power density spectra (PDS; [172]). Most of the power spectral components in the PDS of BHBs are broad and can take the form of a wide power distribution over several decades of frequency or of a more localized peak (quasi-periodic oscillations, QPOs).

QPOs were discovered several decades ago in the X-ray flux emitted from accreting neutron stars and have since been observed in many BHB systems (see, e.g. [116]). It is now clear that QPOs are a common characteristic of accreting BHs and they have been observed also in neutron stars (NS) binaries (e.g. [14, 67, 172]), in cataclysmic variables (see e.g. [129]), in the so-called *ultraluminous X-ray sources* (possibly hosting intermediate-mass BHs or super-Eddington accreting NS, e.g. [8, 165]) and even in active galactic nuclei (AGNs, e.g. [60, 101]).

2.2.2.1 Low-Frequency QPOs

Low-frequency QPOs (LFQPOs) with frequencies ranging from a few MHz to $\sim$30 Hz are a common feature in almost all transient BHBs and were already found in several sources with *Ginga* and divided into different classes (see, e.g. [108] for the case of GX 339-4 and [167] for the case of GS 1124-68). Observations performed with the Rossi X-ray Timing Explorer (RXTE) have led to an extraordinary progress in our knowledge of the properties of variability in BHBs (see [24, 143, 174]): it was only after RXTE was launched that LFQPOs were detected in most observed BHBs (see [173].

Three main types of LFQPOs, dubbed types A, B and C, originally identified in the PDS of XTE J1550-564 (see [69, 142, 178]), have been seen in several sources. The different types of QPOs are currently identified on the basis of their intrinsic properties (mainly centroid frequency and width, but energy dependence and phase lags as well), of the underlying broadband noise components (noise shape and total variability level) and of the relations among these quantities.

- Type-A QPOs (Fig. 2.2, top panel) are characterized by a weak (few per cent rms) and broad ($\nu/\Delta\nu \leq 3$) peak around 6-8 Hz. Neither a subharmonic nor a second harmonic are usually present (possibly because of the width of the fundamental peak), whereas a very low-amplitude red noise is associated with these QPOs. Originally, these LFQPOs were dubbed *type-A-II* by [69]. LFQPOs dubbed *type-A-I* [178] were strong, broad and associated with a very low-amplitude red noise. A shoulder on the right-hand side of this QPO was clearly visible and interpreted as a very broadened second harmonic peak. [35] showed that this *type-A-I* LFQPOs should be classified as a type-B QPOs. Type-A QPOs usually appear in the HSS, just after the transition from the HIMS.

- Type-B QPOs (Fig. 2.2, middle panel) are characterized by a relatively strong (4 % rms) and narrow ($\nu/\Delta\nu \geq 6$) peak, which is found in a narrow range of centroid frequencies around 6 Hz or 1–3 Hz [114]. A weak red noise (few per cent rms or less) is detected at very low frequencies (≤ 0.1 Hz). A weak second harmonic is often present, sometimes together with a subharmonic peak. In a few cases, the subharmonic peak is higher and narrower, resulting in a *cathedral-like* QPO shape (see [34]). Rapid transitions in which type-B LFQPOs appear/disappear are often observed in some sources (e.g. [123]). These transitions are difficult to resolve, as they take place on a timescale shorter than a few seconds. The presence of type-B QPOs essentially defines the SIMS.

- Type-C QPOs (Fig. 2.2, bottom panel) are characterized by a strong (upto 20 % rms), narrow ($\nu/\Delta\nu \geq 10$) and variable peak (its centroid frequency and intensity varying by several per cent in a few days; see, e.g. [116]) at frequencies 0.1–15 Hz, superimposed onto a flat-top noise that steepens above a frequency comparable to that of the QPO. A subharmonic and a second harmonic peak are often seen, and sometimes even a third harmonic peak. The total (QPO plus flat-top noise) fractional rms variability can be as high as 40 %. The frequency of the type-C QPOs correlates both with the flat-top noise break-frequency ([179] and with the characteristic frequency of some broad components seen in the PDS at higher frequency (>20 Hz, see [138]). Type-C QPOs are usually observed during the bright end of the LHS and during the HIMS. In some sources (see e.g. [113, 116]), type-C QPOs can be seen also in the HSS, where they show frequencies that can reach ~ 30 Hz.

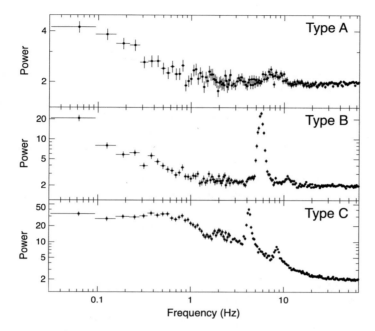

Fig. 2.2 Examples of type-A, type-B and type-C QPOs from our GX 339-4 observations. The contribution of the Poisson noise was not subtracted. Adapted from [114]

2.2.2.2 Models for LFQPOs

Despite LFQPOs being known for several decades, their origin is still not understood and there is no consensus about their physical nature. However, the study of LFQPOs provides an indirect way to explore the accretion flow around black holes (and neutron stars).

The existing models that attempt to explain the origin of LFQPOs are generally based on two different mechanisms: instabilities and geometrical effects. In the latter case, the physical process typically invoked is precession.

Titarchuk and Fiorito [170] proposed the so-called *transition layer model*, where type-C QPOs are the result of viscous magneto-acoustic oscillations of a spherical bounded transition layer, formed by matter from the accretion disc adjusting to the sub-Keplerian boundary conditions near the central compact object.

Cabanac et al. [32] proposed a model to explain simultaneously type-C QPOs and the associated broadband noise. Magneto-acoustic waves propagating within the corona makes it oscillate, causing a modulation in the efficiency of the Comptonization process on the embedded photons. This should produce both the type-C QPOs (thanks to a resonance effect) and the noise that comes with them.

Tagger and Pellat [166] proposed a model based on the *accretion–ejection instability* (AEI), according to which a spiral density wave in the disc, driven by magnetic stresses, becomes unstable by exchanging angular momentum with a Rossby vortex. This instability forms low azimuthal wavenumbers, standing spiral patterns which would be the origin of LFQPOs. Varnière and Tagger [175, 176] suggested that all the tree types of QPOs (A, B and C) can be produced through the AEI in three different regimes: non-relativistic (type-C), relativistic (type-A, where the AEI coexists with the Rossby wave instability (RWI), see [166]) and during the transition between the two regimes (type-B QPOs).

Stella and Vietri [158] proposed the so-called *relativistic precession model* (RPM) to explain the origin and the behaviour of a type of LFQPO (the so-called horizontal-branch oscillation) and of two high-frequency QPOs (the so-called kHz QPOs) in NS X-ray binaries, as the result of the nodal precession, periastron precession and Keplerian motion, respectively, of a self-luminous blob of material in the accretion flow around the compact object. This model was later extended to BHs [159]. Motta et al. [117] recently showed that the RPM provides a good explanation for both type-C QPOs and high-frequency QPO (see below) in at least two BH systems.

Ingram et al. [70] proposed a model based on the relativistic precession as predicted by the theory of general relativity that attempts to explain type-C QPOs and their associated noise. This model requires a cool optically thick, geometrically thin accretion disc [150] truncated at some radius, filled by a hot, geometrically thick accretion flow. This geometry is known as *truncated disc model* [48, 135]. In this framework, type-C QPOs arise from the Lense–Thirring precession of a radially extended section of the hot inner flow that modulates the X-ray flux through a combination of self-occultation, projected area and relativistic effects that become stronger with inclination (see [70]). The broadband noise associated with type-C QPOs, instead, would arise from variations in mass accretion rate from the outer regions of the accretion flow that propagate towards the central compact object, modulating the variations from the inner regions and, consequently, modulating also the radiation in an inclination-independent manner (see [71]).

2.2.2.3   High-Frequency QPOs

Among the most important discoveries of RXTE is the detection of the so-called kHz QPOs in neutron star binaries (see [174]). This result opened a window onto high-frequency phenomena in BHBs. The first observations of the very bright system GRS 1915+105 led to the discovery of a transient oscillation at ∼67 Hz [111], the first high-frequency QPO (HFQPO) in a BHB. Since then, sixteen years of RXTE observations have yielded only a handful of detections in other sources, although GRS 1915+105 seems to be an exception, with a remarkably high number of detected high-frequency QPOs (see e.g. [25]).

The properties of the few confirmed HFQPOs [25] can be summarized as follow:

• They appear only in observations at high-flux/accretion rate. This is at least partly due to a selection effect, but not all high-flux observations lead to the detection of

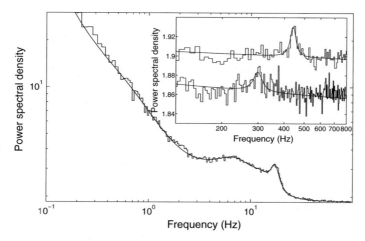

Fig. 2.3 Power spectrum of GRO J1655-40 displaying three simultaneous QPO peaks: the type-C at ~18 Hz, upper and lower high-frequency QPO at ~300 and ~450 Hz, respectively (Fig. 2 from [115])

a HFQPO, all else being equal, indicating that the properties of these oscillations can vary substantially even when all other observables do not change.

- They can be observed as single or double peaks. Only one source, GRS J1655-40 (see Fig. 2.3, showed two clear simultaneous peaks [115, 163], while all the others only showed single peaks, sometimes at different frequencies (see Tab. 1 in [26]. In XTE J1550-564, the two detected peaks [142] have been detected simultaneously after averaging a number of observations, but the lower one with a 2.3 σ significance [103]. Méndez et al. [97], on the basis of their phase lags, suggested that the two detected peaks might be the same physical signal at two different frequencies. H 1743-322 showed a clear HFQPO with a weak second simultaneous peak [66]. A systematic analysis of the data from GRS 1915+105 [22] led to the detection of 51 HFQPOs, most of which at a centroid frequency between 63 and 71 Hz. All detections corresponded to a very limited range in spectral parameters, as measured through hardness ratios. Additional peaks at 27, 34 and 41 Hz were discovered by [16, 22, 164]. The most recent HFQPO discovered, in IGR J17091-3624, is consistent with the average frequency of the 67 Hz QPO in GRS 1915 + 105 [3].
- Typical fractional rms for HFQPOs is 0.5–6 % increasing steeply with energy, in the case of GRS 1915 + 105 reaching more than 19 % at 20–40 keV (see right panel in Fig. 6 of [111]). Quality factors Q^1 are around 5 for the lower peak and 10 for the upper. In GRS 1915 + 105, a typical Q of ~20 is observed, but values as low as 5 and as high as 30 are observed.
- Time lags of HFQPOs have been studied for four sources [97]. The lag spectra of the 67 Hz QPO in GRS 1915 + 105 and IGR J17091-3624 and of the 450 Hz QPO

---

^{1}Q is defined as the ratio between the centroid frequency and FWHM of the QPO peak.

in GRO J1655-40 are hard (hard photon variations lag soft photon variations), while those of the 35 Hz QPO in GRS 1915 + 105 are soft. The 300 Hz QPO in GRO J1655-40 and both HFQPOs in XTE J1550-564 are consistent with zero (suggesting that the two HFQPOs in XTE J1550-564 are the same feature seen at different frequencies).

- For three sources, GRO J1655-40, XTE J1550-564 and XTE J1743-322, the two observed frequencies are close to being in a 3:2 ratio [142, 143, 163], which has led to a family of models, known as *resonance models* (see e.g. [1]). For GRS 1915 + 105, the 67 and 41 Hz QPOs, observed simultaneously, are roughly in 5:3 ratio. The 27 Hz would correspond to 2 in this sequence.

2.2.2.4 Models for HFQPOs

Many models have been proposed to describe HFQPOs of BHBs, all involving in some form the predictions of the theory of general relativity.

The relativistic precession model (RPM) was originally proposed by [158, 160] to explain the origin and the behaviour of the LFQPO and kHz QPOs in NS X-ray binaries and later extended to BHs [115, 117, 159]. The RPM associates three types of QPOs observable in the PDS of accreting compact objects to a combination of the fundamental frequencies of particle motion. The nodal precession frequency (or Lense–Thirring frequency) is associated with type-C QPOs LFQPOs, while the periastron precession frequency and the orbital frequency are associated with the lower and upper HFQPO, respectively (or to the lower and upper kHz QPO in the case of NSs). The relativistic precession model has been proposed in two other versions. In [31], it is assumed that radiation is modulated by the vertical oscillations of a slightly eccentric fluid slender torus formed close to the ISCO. Stuchlik and collaborators proposed a further version of the relativistic precession model that has been studied in many papers by this group. Here, the model is related to the warped disc oscillations discussed by [75] (see below).

The warped disc model proposed by [75, 76] states that the HFQPOs are resonantly excited by specific disc deformation warps. The model was generalized to include precession of the warped disc in [77] and spin-induced perturbations were included in [78].

Abramowicz and Kluźniak [1] and Kluzniak and Abramowicz [79] introduced the nonlinear *resonance model*, which was later studied extensively by them as well as by other authors. This model is based on the assumption that nonlinear 1:2, 1:3 or 2:3 resonance between the orbital and radial epicyclic motion could produce the HFQPOs observed in both BH and NS binaries. Later on, [2] proposed another version of this model, called the Keplerian nonlinear resonance model, where the resonance occurs between the radial epicyclic frequency and the orbital frequency instead of between the radial epicyclic frequency and the vertical frequency. These *resonance models* successfully explain black hole QPOs with frequency ratio consistent with 2:3 or 1:2 (see Sect. 3.2). As a given resonance condition is verified only at a fixed radius in the disc, the QPO frequencies are expected to remain constant, or jump from one resonance to another.

2.2.3 Long-Term Time Evolution

After having defined and discussed the source states in terms of their spectral and timing properties along the HID/HRD/RID diagrams, it is useful to examine the evolution of sources along the HID (and in parallel the other diagrams). A short description can be found in [49], together with an animation.

BHBs spend most of the time in a "quiescent" state, where the accretion rate reaching the central regions of the accretion flow and the black hole is very low, typically lower than $10^{-5}L_{Edd}$ [131, 132], indicated in Fig. 2.4 as QS but actually located below the extension of the Y-axis. Here, the energy spectra indicate that the spectrum deviates from that of the LHS, which hardens as flux decreases. Going down to quiescence, at a level around $10^{-2}L_{Edd}$ the measured power law index starts increasing from 1.5–1.6 until leveling off around 2 at $10^{-5}L_{Edd}$ [131, 181]).

The branch from A to B in Fig. 2.4 corresponds to the LHS and is travelled on timescales which can be as long as months but also so short that the LHS is not observed in pointed observations made on the day following the first alert. We do not

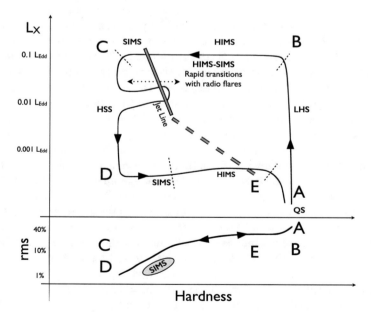

Fig. 2.4 Generic HID and HRD of a black hole binary outburst. The letters refer to main locations described in the text (Fig. 1 from [85])

know directly that the A–B branch was traversed, but we have never seen evidence of the contrary. There are a few cases of systems which never leave the LHS and return to quiescence after having reached a peak (see, e.g. [30] and references therein), not necessarily at low flux [126]). Along this vertical branch, the characteristic frequencies of the strong noise components observed in the PDS increase, while the total fractional rms variability decreases.

Then at B the HIMS is entered. The precise time of the transition can be identified with the changes of timing properties (see [13] and [120]) or through a change in the low-energy properties as was observed in GX 339-4 [66]. The duration of the HIMS, when the source softens from B to the SIMS transition, can be less than a day upto two weeks. A few sources, after entering the HIMS, did not proceed to a further transition and returned to the LHS and then to quiescence, failing to reach a full transition [28, 33, 155]. The characteristic frequencies continue to increase and the total fractional rms to decrease, while an evident type-C QPO appears, also with increasing frequency as the source softens.

If the outburst does not fail, a transition to the SIMS is made. While the energy spectrum below ~ 10 keV is similar to that of the softest HIMS observations, with a hard-component photon index around 2.4, the SIMS is characterized by marked differences in the PDS. In particular, the disappearing of the type-C QPO and the appearing of a type-B QPO at a different frequency are relatively easy to observe. This transition has been seen to take place on a timescale of a few seconds [34]. Multiple back and forth transitions have also been observed, on timescales of days to weeks but also down to minutes [13, 34, 123]. The evolution of the high-energy (> 10 keV) spectral component also changes abruptly here. Around, but not exactly coincident to, this transition, fast discrete relativistic jets are launched, observed either as resolved moving radio spots or as bright radio flares (see [51]). Although it would be tempting to causally associate the disappearance of timing components in the PDS to the ejection of the plasma responsible for them, it has been shown that in some cases the ejection starts a few days *before* the transition [52].

The position of the further transition to the HSS is not easy to identify, as the presence of weak type-A QPOs is not always easy to ascertain. The HSS, after possible back transition to the previous intermediate states, is rather stable and can last for months. The luminosity, which in many systems peaks in the HIMS/SIMS (but see below), decreases, most likely because of a more or less steady decrease in mass accretion rate. A low level of aperiodic time variability is detected, in the form of a power law component in the PDS, with a total fractional rms around 1–2 %.

At a luminosity level well below that of the early HIMS, a new transition takes place (at D). From here on, time is reversed, first the SIMS and then the HIMS are observed, after which the LHS is reached, after which the outburst ends and the source goes back to its quiescent level. However, notice that at low luminosity, the thermal disc contribution is softer, which means that HID points at the same hardness do not correspond to the same energy spectrum. Indeed, the photon index of the hard component at the return SIMS is around 2.1, comparable to that at the start of the first HIMS (point B) [161].

The transition between LHS–HIMS–SIMS–HSS at high flux does not necessarily coincide to the maximum in accretion rate, as it is assumed in the previous description. Indeed, in a few cases the accretion rate continued to increase after the HSS is reached. If this happens, in the HID the source moves up from C moving to the right (see [20, 113]). Both the energy spectrum and the PDS become rather complex [45, 113]).

There are sources for which the diagram appears more complex than this (see examples in [20, 52]), as well as sources which are observed when already in a bright HSS. The latter must have come from quiescence and the data do not exclude that the LHS–HIMS–SIMS branches were followed, only on a much shorter time scale, of the order of the day.

The HID path outlined in Fig. 2.4 shows a clear evidence of hysteresis, as the return path is different from the forward path. In other words, the luminosity (or accretion rate) level at which the hard-to-soft transition takes place is higher than that of the reverse transition [88]. For the prototypical source GX 339-4, which had several outbursts, there is evidence that the higher the first level, the lower the second. A correlation has been found between the waiting time from the previous outburst and the hard X-ray peak, which corresponds to the hard-to-soft transition level (see [182]), but this does not specifically address the issue of hysteresis. The same effect is observed in neutron star transients [88, 119] and implies that luminosity (and hence accretion rate) cannot be an absolute proxy for state. In other words, LHS and HSS can be observed over the same range of luminosities, although the transition from hard to soft does take place at the highest LHS flux (see also [88]). While most generic interpretations for the observed hysteresis invoke the presence of a second parameter in addition to accretion rate, since accretion rate determines the movement along the diagram, but the LHS–HSS transition can take place at different accretion rate levels, [85] associate the hysteresis with the system's memory, in the same way that magnetic hysteresis works. However, what sets the transition level and whether a transition will take place or not is still an open problem. It is interesting to note that while the upper path can take place at very different luminosities, the range spanned by the lower path is much more reduced, around a few percentage of the Eddington luminosity (see [74, 88]). Adding complication, [118] showed that there is a systematic difference in the shape of the HID between high- and low-inclination sources.

A similar hysteresis diagram has been observed in a white dwarf binary, the dwarf nova SS Cyg, using optical emission as soft band and X-ray emission as hard band for the production of the HID [80]. A radio flare was also detected in correspondence to the hard-to-soft transition. Although the comparison is interesting, it has to be noted that in a white dwarf binary the optical/UV emission originates from the accretion disc and the X-ray emission from the boundary layer between the disc and surface. In general sense, what we are observing from these systems is a transition from optically thin to optically thick emission when the density has increased and a reverse transition when the density has decreased, an effect which is also valid for the boundary layer of a dwarf nova (see [84]). The details of the physics of the systems and the transitions do not necessarily need to be the same.

As described in Sect. 2.2.1 and highlighted in more detail by [85], the general outburst evolution (including the basic diagrams, energy spectra and fast time variability) at zero level can be interpreted as a simple evolution of the transition radius within the truncated disc model (see also Sect. 2.5).

2.3 Radio/IR Emission

While the process of accretion onto stellar mass black holes has been studied, mainly in the X-rays, since the beginning of X-ray astronomy, it is only relatively recently that observations at longer wavelengths, notably infrared and radio, have led to the discovery of additional processes such as the ejection of relativistic jets and of wind outflows. It is now clear that without a broadband perspective the complete and complex picture of an accreting compact object in a binary system is impossible to understand. Below, we review the main observational points that in the past decade have led to a major change in perspective.

2.3.1 Radio Jets

Although radio emission from black hole binaries had been detected a long time before (see, e.g. Tananbaum et al. 1971), it was only in the early nineties that coordinated radio campaigns were started, following the discovery of relativistic jets in the radio band. The current observational picture is rather clear and it correlates with the position of the source in the HID (see above). Figure 2.5 shows a sketch of a HID with an indication of the different types of outflow observed.

from below A to B   BHBs spend most of their time in the quiescent state (well below point A in Fig. 2.5), accessible only by the most powerful focusing X-ray telescopes. The X-ray luminosity is $< 10^{33}$ erg/s. Weak flat spectrum radio emission has been detected in this state. As the source brightens in the LHS, the radio luminosity also increases, while the radio spectrum remains flat. In this state, a compact jet has been resolved in a few cases in the radio band (see e.g. [162]). The radio spectrum is flat, consistent with self-absorbed synchrotron emission and extends upto the near infrared, while the polarization level is low. The X-ray flux and radio flux show a strong positive nonlinear correlation (see Sect. 2.3.2.1). The observational data are interpreted with the presence of a compact jet emitting in the radio through synchrotron and moving outwards with moderately relativistic speed ($\Gamma < 2$, see [51]).

from B to C   Point B marks the start of the HIMS. In addition to changes in the rapid variability (total fractional rms), a good marker for this transition is the breakdown of the correlation between the radio-IR flux and X-ray flux (see below and [51, 66]). As the transition to the SIMS is approached, the radio emission generally

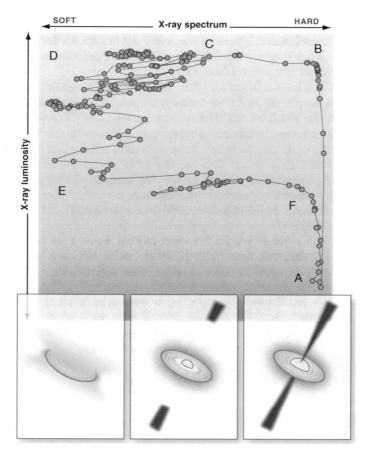

Fig. 2.5 Schematic evolution of a black hole binary along the HID (*top panel*). The three *bottom panels* show the configuration of the system along the *top* branch: where *red* represents radio jets and *yellow* wind outflows. From [49]. Reprinted with permission from AAAS

decreases, with oscillations and small flares [104], while there is indication that the radio spectrum starts steepening [51]. All data indicate that the inner part of the jet undergoes changes in its physical properties.

from C to D   As outlined above, the transition between the HIMS and SIMS is marked by rapid changes in the properties of fast variability, in particular a drop in fractional rms and a switch between the type-C and type-B QPOs. Around (but not precisely in correspondence to) the transition, the jet properties change drastically. One or more large-amplitude flares are observed in the radio (see [52]). When major superluminal ejections are observed, the ejection time can also be traced back to a time close to the transition. Unfortunately, the lack of precise correspondence of the time of jet ejection and that of HIMS–SIMS transition indicates that there cannot be a causal connection. There is indication that these jets have a larger Lorentz factor than the steady jets in the LHS and HIMS, which led to the idea that what is observed is due to internal shocks caused by faster jets hitting the slower components (see [51]).

from D to E   This branch marks the HSS. Until now no radio emission that can be attributed to the central source (and not from ejecta) has been detected down to upper limits of >300 times that of LHS sources at the same X-ray flux (see [147]). Outflows in the form of winds are observed (see 2.4). During this state accretion rate decreases on a long timescale (weeks to months), leading to the vertical track in the HID. However, at high flux there can be multiple transitions to the SIMS/HIMS and back, as measured through the changes in timing properties [13, 20, 34, 49]. Although coverage at lower energies is sparse, the observations are consistent with the presence of weaker radio flares corresponding to the transitions (see e.g. [29, 34]).

from E to F   On the return branch from soft to hard, as we have seen above the reverse track is followed, through the SIMS and the HIMS, on a timescale comparable to that of the upper track. Observations of several transitions from multiple systems have shown that the compact radio jet is re-formed *not* in correspondence with the SIMS–HIMS transition, but when the system has reached the LHS, a delay of several days (see [73, 74] and references therein). At the time of radio reappearance, secondary maxima have been seen in the optical-infrared band, which have been attributed to direct jet emission, although alternative models exist [74].

2.3.2   Accretion–Ejection

2.3.2.1   Correlations

The radio and optical/IR emission from BHBs is attributed to jets (for outflows, see Sect. 2.4) and is closely connected to the accretion properties as measured at higher energies. In particular, the low-energy flux from the compact jet in the LHS shows a strong nonlinear correlation with X-ray flux, modelled with a power law with index $\alpha \sim 0.6$–0.7 (Fig. 2.6, see [39, 58], first discovered in GX 339-4 [40] then extended to other systems [57]). Ignoring the points which appear to follow a different correlation (see below) the scatter in the relation is rather small. Under the assumption that the X-ray flux originates from accretion and therefore is not subject to relativistic beaming like the radio emission from a jet, from the scatter around the correlation [51] have estimated the jet Lorentz factor to be $\Gamma \sim 1$–2. More recently, the correlation has been extended to BHBs in quiescence, notably down to very low fluxes for the first known system A 0620-00 ([56], see also [59]). All stellar mass black holes have likely similar masses (a few to several solar masses). After a correction for mass is added, it was found that the correlation can be extended to active galactic nuclei ([98], see also [132]), spanning 15 orders of magnitude in X-ray flux [56].

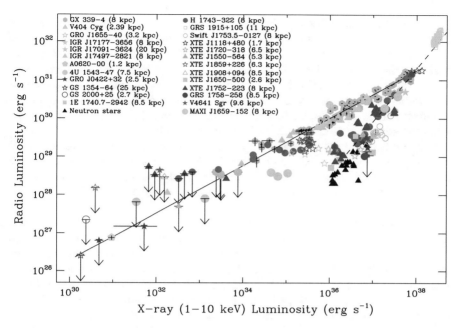

Fig. 2.6 Correlation between the radio and X-ray luminosity for BHBs in the LHS and in quiescence (Fig. 9 from [39])

However, as more multi-wavelength observations became available, the situation has become more complex. A group of sources are found to follow a different correlation, limited to high fluxes and below the main one (often referred to as the "radio-quiet" branch, see [58]). Figure 2.6 shows the clear presence of two branches: the radio-quiet one is steeper, fitted with $\alpha \sim 1$ [58]. Recently, the statistical significance of the presence of two separate correlations has been questioned by [59]. The system H 1743-322 was observed to move from one correlation to the other: during outburst decay, it started on the radio-quiet branch, then as radio luminosity decreased X-ray luminosity stalled until the system reached the radio-loud branch [42]. This clearly indicates that the difference between the two branches cannot be associated with fundamental parameters of the different systems, which would not change during the outburst (see [154]). Coriat et al. [42] explored different possibilities and concluded that the transition between the two branches could be due to a transition from a radiatively efficient to a radiatively inefficient accretion flow, which would make the "radio-quiet" branch an "X-ray loud" one (see [82]). To complicate matters even further, [148] found that during its 2011 outburst MAXI J1836-194 followed a path in the X-radio plane which was significantly steeper and connected the two existing branches. Although the correlation appears to be less *universal* than previously assumed, it is clear that its interpretation can give important insights on both accretion and ejection. For instance, the evaporation/condensation model by

[100] would explain the second, X-ray loud, branch as due to the presence of an inner optically thick accretion disc which would increase the soft photon input available to Comptonization. As this inner disc cannot work at low accretion rate, this interpretation also explains why no lower-branch points are observed at low X-ray fluxes and the switch of H 1743-322. At or after the LHS–HIMS transition, the correlation breaks down (see also below), as the energy spectrum becomes more and more dominated by an optically thick accretion disc and the radio emission also shows non-monotonic variations (see [51]). Interestingly, neutron star LMXBs also show a correlation between the radio and X-ray flux, with three important differences: (1) all NS systems are more radio-quiet at a given X-ray luminosity; (2) the correlation is steeper and (3) the radio emission is not suppressed when the source transits to the soft states [102]. This also points to the presence of a radiatively inefficient regime of accretion for BHBs.

At shorter wavelengths, IR and optical, the situation is complicated by the fact that the accretion disc also contributes to the flux and the jet contribution has to be estimated. In the prototypical system GX 339-4, in the LHS a clear positive correlation between the X-ray and IR flux was found, which terminated abruptly when the source entered the HIMS and the relative contribution between the jet and disc are expected to change [66]. Additional observations have shown that the correlation extends for four orders of magnitude and evidence for a break around $10^{-3}L_{Edd}$ has been found [41]. Recently, a complete analysis of data from 33 systems, both hosting black holes and neutron stars, has been published [146]. For BHBs, they found a correlation with power law index 0.6 extending from quiescence to bright LHS, similar to that observed in the radio.

2.4 Winds and Outflows

Over the last couple of decades we have witnessed the discovery of a multitude of highly ionized absorbers in high-resolution X-ray spectra from both BH and NS X-ray binaries. The first detections were obtained thanks to ASCA on the BH binaries GRO J1655-40 [171] and GRS 1915+105 [83]. Narrow absorption lines in the spectra of these systems identified as Fe XXV and Fe XXVI indicated the first of a myriad of discoveries of photo-ionized plasmas in LMXBs that followed the launch of X-ray observatories such as Chandra, XMM-Newton and Suzaku.

After the first detections, it was soon clear that the photo-ionized absorbers in the form of winds or atmospheres could be ubiquitous to all X-ray binaries (e.g. [127]), but only recently it has become increasingly clear that the presence of these plasmas could be key to our understanding of these systems. It has been suggested that the amount of mass that leaves the system once the plasmas are outflowing can be of the order of or significantly higher than the mass accretion rate transferred through the accretion disc (e.g. [133]), with crucial consequences on the accretion-outflow equilibrium at play in these systems.

Much of the recent work on the accretion/ejection connection at work during the outburst of black hole x-ray transients has focused on the X-ray/radio correlations (see, e.g. [51, 52]). It is now established that transient sources emerge from quiescence entering the low/hard state (see [24]), where a steady radio jet is ubiquitously seen. Then, they rise in luminosity until the hard-to-soft transition takes place. During this transition strong relativistic ejections are often seen—observed as bright radio flares—and connected to the disappearance of the steady radio jets.

Strong disc winds have been predominantly observed in the thermal-dominated high/soft state of BH transients (and most recently in NS systems, see [134]). Conversely, the presence of winds has been excluded by observations in the low/hard state, where no signature of wind-like outflows has been found so far with high level of confidence. Therefore, it seems natural to conclude that in the soft state the radio jets and relativistic ejections seem to be replaced by highly ionized accretion disc winds, that may play a crucial role in the physics of accretion and ejection around compact object. For instance, the effects of the winds on the entire system could be key in the outburst evolution of transient systems, influencing or even triggering the return to the hard state.

2.4.1 Accretion Disc Winds and Atmospheres

To date, several LMXBs, both containing BHs and NSs (see e.g. [43, 133]), have shown narrow absorption lines, more often than not blueshifted (i.e. indicating outflowing material) that have been interpreted as a consequence of the presence of highly ionized material local to the source, opposed to the absorption due to the interstellar medium. Most of these sources are observed at a relatively high inclination angle ($60–70°$), pointing to a distribution of the ionizing plasma close to the accretion disc, with an equatorial or flared geometry.

While the relative depth of the absorption lines detected in the high-resolution spectra of LMXBs allows to determine the column density and the ionization state of the plasma responsible for the lines, their blueshift with respect to their theoretical wavelength provides information on the relative velocity of the plasma. The ionization degree of the plasma is described through the ionization parameter, defined as $\xi = L/nr^2$, where L is the luminosity of the ionizing source, n is the electron density and r is the distance between the ionizing source and the plasma.

The column density of the ionized plasma detected in LMXBs ranges between 10^{21} cm^{-2} and 10^{24} cm^{-2} and there is no obvious difference between the densities measured for BHs and NSs systems. The vast majority of the systems for which $\log(\xi)$ has been measured, show high ionization degrees, with $\log(\xi) > 3$. However, MAXI J1305-70 has shown both high and low ionization degrees, and GX 339-4 (see [151]) has shown only a low ionization degree plasma. Interestingly, among the sources known to show ionization plasmas, GX 339-4 is possibly the one at the lowest inclination: this suggests the possibility that there could be a stratification of the ionized plasma as a function of inclination above the accretion disc. On the other hand, GX 339-4 is affected by relatively low interstellar absorption; thus, the detection of low ionization plasma could be affected by a significant observational bias.

Even though the majority of the ionized plasmas detected so far in LMXBs is characterized by significant intrinsic velocities, for a few sources there is no evidence of an intrinsic plasma velocity. This difference justifies the classification of ionized plasmas in disc winds/outflows and disc atmospheres. When a given system shows absorption lines with significant blueshift and/or a P-Cygni profile, the ionized plasma is classified as *outflow*. If, instead, a given system does not show a significant blueshift in the absorption line, then the ionized plasma takes the name of *atmosphere*. Clearly, the most important difference between the systems shown winds/outflows and the ones showing atmospheres is that only the former loose mass, with a rate that can be even significantly larger than the mass accretion rate (see, e.g. [87, 122, 133]).

Among the sources for which ionized plasmas has been detected, all BH LMXBs shows relatively fast outflows, while only about ∼30 % of NS LMXBs have shown clear winds (see [134]). This difference between NSs and BHS could be due to a significant difference in size of the system, i.e. strong winds are more easily produced in long orbital period (i.e. large systems, and, consequently, large discs). This hypothesis is supported by the fact that the NS systems producing winds are those with orbital periods comparable to the BH ones.

2.4.2   Winds Launching Mechanism

Three main mechanisms have been identified as possible responsibles for the launching of fast winds from the accretion disc of accreting sources: thermal pressure, radiative pressure and magnetic pressure. The mechanism normally invoked to explain the strong winds detected in the radio-free soft states of BH and NS X-ray binaries is the thermal pressure, even though it is important to bear in mind that the dominant wind launching mechanism could change as the system evolves.

2.4.2.1   Thermally Driven Winds

Thermal pressure or Compton heated winds should arise in systems where the accretion disc is irradiated by the central regions of the accretion flow, such as in X-ray binaries or in quasars. The disc gas can be heated to temperatures exceeding 10^7 K mostly through the Compton process, partially evaporating and forming a corona above the disc. This gas, depending on the thermal velocity exceeding or not the local escape velocity, can be either emitted as a thermal wind or stay bound to the disc, forming an atmosphere [12]. The radial extent of this corona only depends on the mass of the central compact object and on the Compton temperature, while it is independent on the luminosity. It has been shown [12] that due to disc rotation a wind can be launched via thermal pressure at radii larger than ∼0.1 r_C, where r_C is the Compton radius, i.e. the radius where the escape velocity equals the isothermal sound speed at the Compton temperature T_C. For very luminous systems (where the radiation force due to electrons must be taken into account), [136] found that strong winds can be launched starting from radii as small as $0.01\ r_{IC}$.

2.4.2.2 Radiatively Driven Winds

Radiation-driven winds might arise from an accretion disc when radiative accelera-tion occurs due to the transfer of momentum from the photon field to the corona that leaves the disc in the form of an outflow. Assuming that the wind has optical depth τ_e higher than 1 and completely surrounds a source of radiation with luminosity L_{bol}, the multiple scattering of photons within the wind will lead to a wind momentum that exceed the photon momentum of the primary emission. In most sources, however, the material constituting the wind appears to be far too ionized to allow an effec-tive momentum transfer and a consequent launching mechanism. In other words, the momentum-flux in radiatively driven winds cannot exceed that of the radiation field, even considering the effects of radiation of free electrons (see [144]).

2.4.2.3 Magnetically Driven Winds

Disc winds can also be driven by magnetic forces. These winds are expected to be centrifugally accelerated down open, rotating, poloidal magnetic fields anchored in the disc (see [27]). Being these winds accelerated by the effect of magnetic torques from magnetic fields embedded in the accretion disc, there must be an intimate connection between the mass loss in the wind and the accretion onto the black hole. According to the theory of MHD winds, material is centrifugally lifted off the disc at a certain launching radius R_L and is continuously accelerated by magnetic forces until the flow becomes super-Alfvenic at a radius $R_A > R_L$. Accurate models suggest that the ratio R_A/R_L ranges between 2 and 3 (see [139]). Reynolds [144] has shown that MHD torques would be able to produce Compton thick winds only if (1) the accretion rate is a significant fraction of the Eddington rate, (2) the radiative efficiency is low and (3) the Alfven radius is very close to the launching radius ($R_A \sim R_L$). Thus, MHD driving could become a viable explanation for disc winds only in a very limited number of cases.

2.5 The Full Accretion–Ejection Picture

Despite the number of complications and additional details, the overall picture is now much more clear than before the RXTE data. Not all transient sources behave in the fundamental diagrams as neatly as the one shown in Fig. 2.1, but the general pattern appears to be followed by most sources and this regularity points towards fundamental aspects of accretion onto black holes (see [52]). Indeed, a comparison with different classes of related sources shows that this must be the case.

Persistent BHBs. A few persistent systems are known in our galaxy and the Magellanic Clouds. Some of them appear locked in a single state. LMC X-1 has always been observed in the HSS (see [145]), as the bright galactic source 4U 1957 + 11 [124]. The second BHB in the LMC, LMC X-3, was only observed in the HSS until with the extensive coverage of RXTE brief transitions to and from the LHS were discovered [152]. The source GRS 1758-258 in the galactic centre region is mostly in the LHS, but shows sporadic transitions to the HSS (see [157]). The bright source 4U 1755-33 was very bright and in the HSS until 1996, when it went into quiescence (see [5]), but unfortunately the decay into quiescence was not covered by observations. The brightest and best known Cyg X-1, the first black hole candidate, is usually found in the LHS and makes rather frequent transitions to the HIMS, to rarely reach the HSS when radio emission is observed to drop (see [63] and references therein). A HID from the RXTE observations of Cyg X-1 is shown in [20]. Overall, none of these systems shows a behaviour inconsistent with the above picture, although clearly none of them shows a full "transient" cycle. In particular, the LHS–HIMS and HMIS–LHS transitions of Cyg X-1 do not show any sign of hysteresis. Interestingly, corresponding to one of the softest events of Cyg X-1, a radio flare was observed, compatible with a jet ejection in correspondence to a transition to the HSS [180].

A case of its own is represented by GRS 1915 + 105, which has started an outburst in 1992 and is still active at the time of writing. Its behaviour is very different from all other systems, although a few sources have been found to match it rather precisely for some limited time [4, 9]. The original idea was that this peculiar behaviour is connected to a very high accretion rate, which would put the source above the standard "q" diagram (see [20] for an HID of state-C, i.e. hard, intervals of GRS 1915 + 105). However, recently the same type of structured variability has been found in the Rapid Burster, a neutron star binary, at luminosities well below the Eddington limit, casting doubts on this interpretation [9]. At any rate, the short-term variability of GRS 1915 + 105 is not different from those observed in other BHBs: during state-C, whose energy spectrum corresponds to a LHS/HIMS, the PDS is a typical LHS/HIMS [20, 140, 141], during softer and brighter states (called A and B, see [15] the PDS is similar to that of "anomalous" states at high accretion rate see in other BHBs [20, 140]. HFQPOs are observed only in a very narrow region of the HID [23]. Evidence for a type-B QPO during fast transition has also been presented [153]. What is peculiar here is the structure of alternating states.

Neutron star binaries. Neutron star low-mass X-ray binaries show many similarities in their emission properties with BHBs. Their detailed X-ray energy spectra are rather different as the component of direct emission from the surface of the compact object and the boundary layer is very bright. However, it is now clear that the properties of fast time variability can be strongly connected to those of BHBs (see [18, 35, 125, 138, 179]. The first evidence of a hysteresis pattern in the HID was presented from the low-luminosity persistent system 4U 1636-53 [14]. "q" diagrams were shown for the outbursts of Aql X-1 [20, 81]. More recently, a full analysis of

Fig. 2.7 RID diagram with the regions occupied by BHBs and different classes of NS LMXBs (Z: Z sources, BA: bright atoll sources, the *grey* "q" corresponds to atoll sources; Fig. 10 from [119]). The X-axis in this diagram is fractional rms variability, which correlates almost linearly with hardness, which means a HID would look almost identical

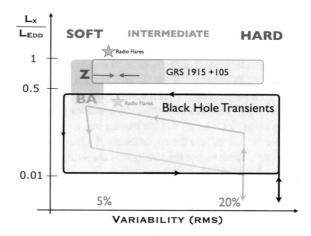

RXTE data has shown that the observed pattern is very similar and a strong connection has been drawn [119]. As the difference between the classes of neutron star LMXBs has been finally attributed to accretion rate levels (see [68], a complete diagram for NS and BH binaries could be produced (see Fig. 2.7). A strong connection with the BHBs was found in the radio emission of Aql X-1, completing the picture [81]. The radio-X correlation for NS binaries lies below that of BHBs, suggesting a difference in the emission efficiency between the two classes of systems in their hard state [102].

ULX. In the case of ultraluminous X-ray sources (ULX), the comparison is not simple, given the lower statistics and sparser coverage of observations, not to mention the fact that there is no agreement on the nature of the systems (see [53] for a review). There are similarities in the timing properties, but there is no consensus on the interpretation of the few QPOs detected in ULX in terms of those in BHBs (see, e.g. [121, 128, 165]). The energy spectra are even more complex and only basic comparisons can be made, not always in agreement with each other [21, 156]. State transitions have been observed (see [72] and references therein), but only with few observations. When standard diagrams have been produced for a sample of ULX, the results were also complex and difficult to interpret [130]. In the case of the hyperluminous source ESO 243-49 HLX-1, a number of XMM-Newton and Chandra observations were used to produce HID and HRD, showing compatibility with the BHB diagrams [149]. Radio flares have been observed from this source, although the lack of high-SNR X-ray coverage prevents a precise association with source states [177].

AGN. The association between the AGN properties and BHB states is also complex. While Quasars and Blazars are thought to be jet-dominated systems and Seyfert II AGN suffer of large absorption, a comparison can be attempted with Seyfert I systems. On the spectral side, the energy distribution in X-rays is similar to that of galactic systems in the LHS, although for AGN the thermal accretion disc component is not in the "standard" X-ray band. Indeed, the production of a HID from RXTE data of Sy I objects shows that almost all of them lie on the LHS branch. However, analysis of variability, which in the case of AGN must be observed upto time scales of years, shows that their PDS is more similar to that of HSS binaries (see, e.g. [94, 96]). Two systems stand out both spectrally and in timing properties: Ton S180 and Ark 564. When placed on a standard HID, renormalized for flux differences, these object lie in an intermediate-hardness area [20]. The exact connection to the BHB diagram is not easy as the horizontal branches are caused by both a softening of the LHS component and the appearance of a thermal disc component, absent in the AGN 2–10 keV spectra, but the position is certainly intermediate. On the timing side, these two objects also show properties which can be identified of more intermediate states [6, 94, 95].

This general behaviour must be understood in terms of basic parameters. Interpretations of the overall evolution in the HID have been proposed [10, 11, 88]. However, the correlated properties of the emission of relativistic jets and winds from the system must be included in order to reach a global understanding. This of course complicates the problem and only basic attempts have been made [49, 51]. A zero-level approach has been proposed recently, which in addition to the X-ray properties incorporates a model for the production of poloidal magnetic fields, which can be a crucial ingredient for the production of jets [85]. The "cosmic battery" scenario (see [36–38, 86]) can be the link between the properties of the accretion flow and generation of relativistic jets, which is worth exploring further.

2.6   Conclusions

Thanks to the availability of high-quality data from past and current high-energy missions, our knowledge and understanding of black hole transients has increased significantly in the past two decades. Instrumental to this advancements have been observations at other wavelengths, which have allowed us to study physical components, such as relativistic jets and outflows, that were completely ignored before. It is now clear that studying the emission over a very broad range of energies is the only way to properly characterize the physical properties of these objects. For the study of fast X-ray time variability, the gap left by the demise of RossiXTE has just been filled by the launch of the Indian satellite Astrosat, which at the time of writing is still in the performance verification phase. Theoretical modelling of observational data is now turning to the interpretation of joint spectral-timing properties, a crucial step to study rapidly varying phenomena. The field is evolving very rapidly: however, while the amount of available details make the full picture more complicated, the basic patterns are converging towards a solid set of observational points that must be at the base of all theoretical models. Above, we have focused onto those patterns with the aim of outlining general properties of black hole transients. This is by no means complete (see Middleton's chapter on detailed spectral models), but it should constitute a solid starting point upon which layers of complexity can later be laid.

References

1. M.A. Abramowicz, W. Kluźniak, A precise determination of black hole spin in GRO J1655–40. Astron. Astrophys. **374**, L19–L20 (2001)
2. M.A. Abramowicz, W. Kluzniak, Z. Stuchlik, G. Torok, The orbital resonance model for twin peak khz QPOs. Astrophysics, arXiv: e-prints (2004)
3. D. Altamirano, T. Belloni, Discovery of high-frequency quasi-periodic oscillations in the black hole candidate IGR J17091–3624. ApJ **747**, L4 (2012)
4. D. Altamirano, T. Belloni, M. Linares, M. van der Klis, R. Wijnands, P.A. Curran, M. Kalamkar, H. Stiele, S. Motta, T. Muñoz-Darias, P. Casella, H. Krimm, The faint "Heartbeats" of IGR J17091–3624: an exceptional black hole candidate. ApJ **742**, L17 (2011)
5. L. Angelini, N.E. White, An XMM-newton observation of 4U 1755–33 in quiescence: evidence of a fossil X-ray jet. ApJ **586**, L71–L75 (2003)
6. P. Arévalo, I.E. Papadakis, P. Uttley, I.M. McHardy, W. Brinkmann, Spectral-timing evidence for a very high state in the narrow-line Seyfert 1 Ark 564. MNRAS **372**, 401–409 (2006)
7. M.A. Padilla, N. Degenaar, A. Patruno, D.M. Russell, M. Linares, T.J. Maccarone, J. Homan, R. Wijnands, X-ray softening in the new X-ray transient XTE J1719-291 during its 2008 outburst decay. MNRAS, **417**, 659–665 (2011)
8. M. Bachetti, F.A. Harrison, D.J. Walton, B.W. Grefenstette, D. Chakrabarty, F. Fürst, D. Barret, A. Beloborodov, S.E. Boggs, F.E. Christensen, W.W. Craig, A.C. Fabian, C.J. Hailey, A. Hornschemeier, V. Kaspi, S.R. Kulkarni, T. Maccarone, J.M. Miller, V. Rana, D. Stern, S.P. Tendulkar, J. Tomsick, N.A. Webb, W.W. Zhang, An ultraluminous X-ray source powered by an accreting neutron star. Nature **514**, 202–204 (2014)
9. T. Bagnoli, J.J.M. in't Zand, Discovery of GRS 1915+105 variability patterns in the Rapid Burster. MNRAS, **450**, L52–L56 (2015)
10. M.C. Begelman, P.J. Armitage, A mechanism for hysteresis in black hole binary state transitions. ApJ **782**, L18 (2014)
11. M.C. Begelman, P.J. Armitage, C.S. Reynolds, Accretion disk dynamo as the trigger for X-ray binary state transitions. ApJ **809**, 118 (2015)
12. M.C. Begelman, C.F. McKee, G.A. Shields, Compton heated winds and coronae above accretion disks I dynamics. ApJ **271**, 70–88 (1983)
13. T. Belloni, J. Homan, P. Casella, M. van der Klis, E. Nespoli, W.H.G. Lewin, J.M. Miller, M. Méndez, The evolution of the timing properties of the black-hole transient GX 339–4 during its 2002/2003 outburst. Astron. Astrophys. **440**, 207–222 (2005)
14. T. Belloni, J. Homan, S. Motta, E. Ratti, M. Méndez, Rossi XTE monitoring of 4U1636-53 - I. Long-term evolution and kHz quasi-periodic oscillations. MNRAS **379**, 247–252 (2007)
15. T. Belloni, M. Klein-Wolt, M. Méndez, M. van der Klis, J. van Paradijs, A model-independent analysis of the variability of GRS 1915+105. Astron. Astrophys. **355**, 271–290 (2000)
16. T. Belloni, M. Méndez, C. Sánchez-Fernández, The high-frequency QPOs in GRS 1915+105. Astron. Astrophys. **372**, 551–556 (2001)
17. T. Belloni, M. Mendez, M. van der Klis, G. Hasinger, W.H.G. Lewin, J. van Paradijs, An intermediate state of Cygnus X-1. ApJ **472**, L107 (1996)
18. T. Belloni, D. Psaltis, M. van der Klis, A unified description of the timing features of accreting X-ray binaries. ApJ **572**, 392–406 (2002)
19. T. Belloni, M. van der Klis, W.H.G. Lewin, J. van Paradijs, T. Dotani, K. Mitsuda, S. Miyamoto, Energy dependence in the quasi-periodic oscillations and noise of black hole candidates in the very high state. Astron. Astrophys. **322**, 857–867 (1997)

20. T.M. Belloni, *States and Transitions in Black Hole Binaries*. Lecture Notes in Physics, vol. 794 (Springer, Heidelberg, 2010), p. 53. ISBN 978-3-540-76936-1
21. T.M. Belloni, Black-hole states in external galaxies. Astronomische Nachrichten **332**, 324 (2011)
22. T.M. Belloni, D. Altamirano, Discovery of a 34 Hz quasi-periodic oscillation in the X-ray emission of GRS 1915+105. MNRAS **432**, 19–22 (2013)
23. T.M. Belloni, D. Altamirano, High-frequency quasi-periodic oscillations from GRS 1915+105. MNRAS **432**, 10–18 (2013)
24. T.M. Belloni, S.E. Motta, T. Muñoz-Darias, Black hole transients. Bull. Astron. Soc. India **39**, 409–428 (2011)
25. T.M. Belloni, A. Sanna, M. Méndez, High-frequency quasi-periodic oscillations in black hole binaries. MNRAS **426**, 1701–1709 (2012). November
26. T.M. Belloni, L. Stella, Fast variability from black-hole binaries. Space Sci. Rev. **183**, 43–60 (2014)
27. R.D. Blandford, D.G. Payne, Hydromagnetic flows from accretion discs and the production of radio jets. MNRAS **199**, 883–903 (1982)
28. C. Brocksopp, R.M. Bandyopadhyay, R.P. Fender, "Soft X-ray transient" outbursts which are not soft. New Astron. **9**, 249–264 (2004)
29. C. Brocksopp, R.P. Fender, M. McCollough, G.G. Pooley, M.P. Rupen, R.M. Hjellming, C.J. de la Force, R.E. Spencer, T.W.B. Muxlow, S.T. Garrington, S. Trushkin, Initial low/hard state, multiple jet ejections and X-ray/radio correlations during the outburst of XTE J1859+226. MNRAS **331**, 765–775 (2002)
30. C. Brocksopp, P.G. Jonker, D. Maitra, H.A. Krimm, G.G. Pooley, G. Ramsay, C. Zurita, Disentangling jet and disc emission from the 2005 outburst of XTE J1118+480. MNRAS **404**, 908–916 (2010)
31. M. Bursa, High-frequency QPOs in GRO J1655-40: Constraints on resonance models by spectral fits, in *RAGtime 6/7: workshops on black holes and neutron stars*, eds. by S. Hledík, Z. Stuchlík (2005). pp. 39–45
32. C. Cabanac, G. Henri, P.-O. Petrucci, J. Malzac, J. Ferreira, T.M. Belloni, Variability of X-ray binaries from an oscillating hot corona. MNRAS **404**, 738–748 (2010)
33. F. Capitanio, T. Belloni, M. Del Santo, P. Ubertini, A failed outburst of H1743–322. MNRAS **398**, 1194–1200 (2009)
34. P. Casella, T. Belloni, J. Homan, L. Stella, A study of the low-frequency quasi-periodic oscillations in the X-ray light curves of the black hole candidate ¡ASTROBJ¿ XTE J1859+226¡/ASTROBJ¿. Astron. Astrophys. **426**, 587–600 (2004)
35. P. Casella, T. Belloni, L. Stella, The ABC of low-frequency quasi-periodic oscillations in black hole candidates: analogies with Z sources. ApJ **629**, 403–407 (2005)
36. I. Contopoulos, D. Kazanas, A cosmic battery. ApJ **508**, 859–863 (1998)
37. I. Contopoulos, A. Nathanail, M. Katsanikas, The cosmic battery in astrophysical accretion disks. ApJ **805**, 105 (2015)
38. I. Contopoulos, D.B. Papadopoulos, The cosmic battery and the inner edge of the accretion disc. MNRAS **425**, 147–152 (2012)
39. S. Corbel, M. Coriat, C. Brocksopp, A.K. Tzioumis, R.P. Fender, J.A. Tomsick, M.M. Buxton, C.D. Bailyn, The 'universal' radio/X-ray flux correlation: the case study of the black hole GX 339–4. MNRAS **428**, 2500–2515 (2013)
40. S. Corbel, M.A. Nowak, R.P. Fender, A.K. Tzioumis, S. Markoff, Radio/X-ray correlation in the low/hard state of GX 339–4. Astron. Astrophys. **400**, 1007–1012 (2003)
41. M. Coriat, S. Corbel, M.M. Buxton, C.D. Bailyn, J.A. Tomsick, E. Körding, E. Kalemci, The infrared/X-ray correlation of GX 339–4: probing hard X-ray emission in accreting black holes. MNRAS **400**, 123–133 (2009)
42. M. Coriat, S. Corbel, L. Prat, J.C.A. Miller-Jones, D. Cseh, A.K. Tzioumis, C. Brocksopp, J. Rodriguez, R.P. Fender, G.R. Sivakoff, Radiatively efficient accreting black holes in the hard state: the case study of H1743–322. MNRAS **414**, 677–690 (2011)

43. M.D. Trigo, A.N. Parmar, L. Boirin, M. Méndez, J.S. Kaastra, Spectral changes during dipping in low-mass X-ray binaries due to highly-ionized absorbers. Astron. Astrophys. **445**, 179–195 (2006)
44. C. Done, M. Gierlinski, A. Kubota, Modelling the behaviour of accretion flows in X-ray binaries. Everything you always wanted to know about accretion but were afraid to ask. Astron. Astrophys. **15**, 1–66 (2007)
45. C. Done, A. Kubota, Disc-corona energetics in the very high state of Galactic black holes. MNRAS **371**, 1216–1230 (2006)
46. R.J.H. Dunn, R.P. Fender, E.G. Körding, T. Belloni, C. Cabanac, A global spectral study of black hole X-ray binaries. MNRAS **403**, 61–82 (2010)
47. M. Elvis, C.G. Page, K.A. Pounds, M.J. Ricketts, M.J.L. Turner, Discovery of powerful transient X-ray source A0620-00 with ariel V sky survey experiment. Nature, **257**, 656 (1975)
48. A.A. Esin, J.E. McClintock, R. Narayan, Advection-dominated accretion and the spectral states of black hole X-ray binaries: application to nova MUSCAE 1991. ApJ **489**, 865 (1997)
49. R. Fender, T. Belloni, Stellar-mass black holes and ultraluminous X-ray sources. Science **337**, 540 (2012)
50. R. Fender, E. Gallo, An overview of jets and outflows in stellar mass black holes. Space Sci. Rev. **183**, 323–337 (2014)
51. R.P. Fender, T.M. Belloni, E. Gallo, Towards a unified model for black hole X-ray binary jets. MNRAS **355**, 1105–1118 (2004)
52. R.P. Fender, J. Homan, T.M. Belloni, Jets from black hole X-ray binaries: testing, refining and extending empirical models for the coupling to X-rays. MNRAS **396**, 1370–1382 (2009)
53. H. Feng, R. Soria, Ultraluminous X-ray sources in the Chandra and XMM-Newton era. New Astron. Rev. **55**, 166–183 (2011)
54. J. Ferreira, P.-O. Petrucci, G. Henri, L. Saugé, G. Pelletier, A unified accretion-ejection paradigm for black hole X-ray binaries I. The dynamical constituents. Astron. Astrophys. **447**, 813–825 (2006)
55. E. Gallo, *Radio Emission and Jets From Microquasars*, in ed. by T. Belloni, Lecture Notes in Physics, vol. 794 (Springer, Berlin, 2010), p. 85
56. E. Gallo, R.P. Fender, J.C.A. Miller-Jones, A. Merloni, P.G. Jonker, S. Heinz, T.J. Maccarone, M. van der Klis, A radio-emitting outflow in the quiescent state of A0620–00: implications for modelling low-luminosity black hole binaries. MNRAS **370**, 1351–1360 (2006)
57. E. Gallo, R.P. Fender, G.G. Pooley, A universal radio-X-ray correlation in low/hard state black hole binaries. MNRAS **344**, 60–72 (2003)
58. E. Gallo, B.P. Miller, R. Fender, Assessing luminosity correlations via cluster analysis: evidence for dual tracks in the radio/X-ray domain of black hole X-ray binaries. MNRAS **423**, 590–599 (2012)
59. E. Gallo, J.C.A. Miller-Jones, D.M. Russell, P.G. Jonker, J. Homan, R.M. Plotkin, S. Markoff, B.P. Miller, S. Corbel, R.P. Fender, The radio/X-ray domain of black hole X-ray binaries at the lowest radio luminosities. MNRAS **445**, 290–300 (2014)
60. M. Gierliński, M. Middleton, M. Ward, C. Done, A periodicity of 1hour in X-ray emission from the active galaxy RE J1034+396. Nature **455**, 369–371 (2008)
61. M. Gierliński, A.A. Zdziarski, J. Poutanen, P.S. Coppi, K. Ebisawa, W.N. Johnson, Radiation mechanisms and geometry of Cygnus X-1 in the soft state. MNRAS **309**, 496–512 (1999)
62. M. Gilfanov, X-ray emission from black-hole binaries, in *The Jet Paradigm, Lecture Notes in Physics, vol. 794 (Springer, Heidelberg, 2010), p. 17. ISBN 978-3-540-76936-1*
63. V. Grinberg, N. Hell, K. Pottschmidt, M. Böck, M.A. Nowak, J. Rodriguez, A. Bodaghee, M.C. Bel, G.L. Case, M. Hanke, M. Kühnel, S.B. Markoff, G.G. Pooley, R.E. Rothschild, J.A. Tomsick, C.A. Wilson-Hodge, J. Wilms, Long term variability of Cygnus X-1. V. State definitions with all sky monitors. Astron. Astrophys. **554**, A88 (2013)
64. F. Haardt, L. Maraschi, X-ray spectra from two-phase accretion disks. ApJ **413**, 507–517 (1993)
65. J. Homan, T. Belloni, The evolution of black hole states. Astrophys. Space Sci. **300**, 107–117 (2005)

66. J. Homan, M. Buxton, S. Markoff, C.D. Bailyn, E. Nespoli, T. Belloni, Multiwavelength observations of the 2002 outburst of GX 339–4: two patterns of X-ray-optical/near-infrared behavior. ApJ **624**, 295–306 (2005)
67. J. Homan, M. van der Klis, P.G. Jonker, R. Wijnands, E. Kuulkers, M. Méndez, W.H.G. Lewin, RXTE observations of the neutron star low-mass X-ray binary GX 17+2: correlated X-ray spectral and timing behavior. ApJ **568**, 878–900 (2002)
68. J. Homan, M. van der Klis, R. Wijnands, T. Belloni, R. Fender, M. Klein-Wolt, P. Casella, M. Méndez, E. Gallo, W.H.G. Lewin, N. Gehrels, Rossi X-ray timing explorer observations of the first transient Z source XTE J1701–462: shedding new light on mass accretion in luminous neutron star X-ray binaries. ApJ **656**, 420–430 (2007)
69. J. Homan, R. Wijnands, M. van der Klis, T. Belloni, J. van Paradijs, M. Klein-Wolt, R. Fender, M. Méndez, Correlated X-ray spectral and timing behavior of the black hole candidate XTE J1550–564: a new interpretation of black hole states. ApJs **132**, 377–402 (2001)
70. A. Ingram, C. Done, P.C. Fragile, Low-frequency quasi-periodic oscillations spectra and Lense-Thirring precession. MNRAS **397**, L101–L105 (2009)
71. A. Ingram, M. van der Klis, An exact analytic treatment of propagating mass accretion rate fluctuations in X-ray binaries. MNRAS **434**, 1476–1485 (2013)
72. P. Kaaret, H. Feng, A state transition of the luminous X-ray binary in the low-metallicity blue compact dwarf galaxy I Zw 18. ApJ **770**, 20 (2013)
73. E. Kalemci, M.Ö. Arabacı, T. Güver, D.M. Russell, J.A. Tomsick, J. Wilms, G. Weidenspointner, E. Kuulkers, M. Falanga, T. Dinçer, S. Drave, T. Belloni, M. Coriat, F. Lewis, T. Muñoz-Darias, Multiwavelength observations of the black hole transient Swift J1745–26 during the outburst decay. MNRAS **445**, 1288–1298 (2014)
74. E. Kalemci, T. Dinçer, J.A. Tomsick, M.M. Buxton, C.D. Bailyn, Y.Y. Chun, Complete multiwavelength evolution of galactic black hole transients during outburst decay. I. Conditions for "Compact" jet formation. ApJ **779**, 95 (2013)
75. S. Kato, Resonant excitation of disk oscillations by warps: a model of kHz QPOs. PASJ **56**, 905–922 (2004)
76. S. Kato, Wave-warp resonant interactions in relativistic disks and kHz QPOs. PASJ **56**, 559–567 (2004)
77. S. Kato, A resonance model of quasi-periodic oscillations of low-mass X-ray binaries. PASJ **57**, L17–L20 (2005)
78. S. Kato, Quasi-periodic oscillations resonantly induced on spin-induced deformed-disks of neutron stars. PASJ **57**, 679–690 (2005)
79. W. Kluzniak, M.A. Abramowicz, The physics of kHz QPOs—strong gravity's coupled anharmonic oscillators. Astrophysics, ArXiv: e-prints (2001)
80. E. Körding, M. Rupen, C. Knigge, R. Fender, V. Dhawan, M. Templeton, T. Muxlow, A transient radio jet in an erupting dwarf nova. Science **320**, 1318 (2008)
81. E.G. Körding, Common disc-jet coupling in accreting objects. Astrophys. Space Sci. **311**, 143–147 (2007)
82. E.G. Körding, R.P. Fender, S. Migliari, Jet-dominated advective systems: radio and X-ray luminosity dependence on the accretion rate. MNRAS **369**, 1451–1458 (2006)
83. T. Kotani, K. Ebisawa, T. Dotani, H. Inoue, F. Nagase, Y. Tanaka, Y. Ueda, ASCA observations of the absorption line features from the superluminal jet source GRS 1915+105. ApJ **539**, 413–423 (2000)
84. E. Kuulkers, A. Norton, A. Schwope, B. Warner, *X-rays from Cataclysmic Variables* (2006), pp. 421–460
85. N.D. Kylafis, T.M. Belloni, Accretion and ejection in black-hole X-ray transients. Astron. Astrophys. **574**, A133 (2015)
86. N.D. Kylafis, I. Contopoulos, D. Kazanas, D.M. Christodoulou, Formation and destruction of jets in X-ray binaries. Astron. Astrophys. **538**, A5 (2012)
87. J.C. Lee, C.S. Reynolds, R. Remillard, N.S. Schulz, E.G. Blackman, A.C. Fabian, High-resolution chandra HETGS and rossi X-ray timing explorer observations of GRS 1915+105: a hot disk atmosphere and cold gas enriched in iron and silicon. ApJ **567**, 1102–1111 (2002)

88. T.J. Maccarone, P.S. Coppi, Hysteresis in the light curves of soft X-ray transients. MNRAS **338**, 189–196 (2003)

89. J. Malzac, A.M. Dumont, M. Mouchet, Full radiative coupling in two-phase models for accreting black holes. Astron. Astrophys. **430**, 761–769 (2005)

90. S. Markoff, From multiwavelength to mass scaling: accretion and ejection in microquasars and AGN, in ed. by T. Belloni, Lecture Notes in Physics, vol. 794 (Springer, Berlin, 2010), p. 143

91. S. Markoff, H. Falcke, R. Fender, A jet model for the broadband spectrum of XTE J1118+480. Synchrotron emission from radio to X-rays in the Low/Hard spectral state. Astron. Astrophys. **372**, L25–L28 (2001)

92. C.B. Markwardt, J.H. Swank, New outburst of GRO J1655–40? Astron. Telegr. **414**, 1 (2005)

93. J.E. McClintock, R.A. Remillard, The black hole binary A0620–00. ApJ **308**, 110–122 (1986)

94. I. McHardy, X-ray variability of AGN and relationship to galactic black hole binary systems, in ed. by T. Belloni, Lecture Notes in Physics, vol. 794 (Springer, Berlin, 2010), p. 203

95. I.M. McHardy, P. Arévalo, P. Uttley, I.E. Papadakis, D.P. Summons, W. Brinkmann, M.J. Page, Discovery of multiple Lorentzian components in the X-ray timing properties of the narrow line seyfert 1 Ark 564. MNRAS **382**, 985–994 (2007)

96. I.M. McHardy, K.F. Gunn, P. Uttley, M.R. Goad, MCG-6-30-15: long time-scale X-ray variability, black hole mass and active galactic nuclei high states. MNRAS **359**, 1469–1480 (2005)

97. M. Méndez, D. Altamirano, T. Belloni, A. Sanna, The phase lags of high-frequency quasi-periodic oscillations in four black hole candidates. MNRAS (2013)

98. A. Merloni, S. Heinz, T. di Matteo, A fundamental plane of black hole activity. MNRAS **345**, 1057–1076 (2003)

99. F. Meyer, E. Meyer-Hofmeister, Accretion disk evaporation by a coronal siphon flow. Astron. Astrophys. **288**, 175–182 (1994)

100. E. Meyer-Hofmeister, F. Meyer, The relation between radio and X-ray luminosity of black hole binaries: affected by inner cool disks? Astron. Astrophys. **562**, A142 (2014)

101. M. Middleton, C. Done, The X-ray binary analogy to the first AGN quasi-periodic oscillation. MNRAS **403**, 9–16 (2010)

102. S. Migliari, R.P. Fender, Jets in neutron star X-ray binaries: a comparison with black holes. MNRAS **366**, 79–91 (2006)

103. J.M. Miller, R. Wijnands, J. Homan, T. Belloni, D. Pooley, S. Corbel, C. Kouveliotou, M. van der Klis, W.H.G. Lewin, High-frequency quasi-periodic oscillations in the 2000 outburst of the Galactic microquasar XTE J1550–564. ApJ **563**, 928–933 (2001)

104. J.C.A. Miller-Jones, R.P. Fender, E. Nakar, Opening angles, Lorentz factors and confinement of X-ray binary jets. MNRAS **367**, 1432–1440 (2006)

105. I.F. Mirabel, L.F. Rodríguez, A superluminal source in the Galaxy. Nature **371**, 46–48 (1994)

106. I.F. Mirabel, L.F. Rodríguez, B. Cordier, J. Paul, F. Lebrun, A double-sided radio jet from the compact Galactic centre annihilator 1E1740.7-2942. Nature **358**, 215–217 (1992)

107. S. Miyamoto, S. Iga, S. Kitamoto, Y. Kamado, Another canonical time variation of X-rays from black hole candidates in the very high flare state? ApJ **403**, L39–L42 (1993)

108. S. Miyamoto, S. Kitamoto, A jet model for a very high state of GX 339–4. ApJ **374**, 741–743 (1991)

109. S. Miyamoto, S. Kitamoto, K. Hayashida, W. Egoshi, Large hysteretic behavior of stellar black hole candidate X-ray binaries. ApJ **442**, L13–L16 (1995)

110. S. Miyamoto, S. Kitamoto, S. Iga, H. Negoro, K. Terada, Canonical time variations of X-rays from black hole candidates in the low-intensity state. ApJ **391**, L21–L24 (1992)

111. E.H. Morgan, R.A. Remillard, J. Greiner, RXTE observations of QPOs in the black hole candidate GRS 1915+105. ApJ **482**, 993 (1997)

112. S. Motta, T. Belloni, J. Homan, The evolution of the high-energy cut-off in the X-ray spectrum of GX 339–4 across a hard-to-soft transition. MNRAS **400**, 1603–1612 (2009)

113. S. Motta, J. Homan, T. Muñoz-Darias, P. Casella, T.M. Belloni, B. Hiemstra, M. Méndez, Discovery of two simultaneous non-harmonically related quasi-periodic oscillations in the 2005 outburst of the black hole binary GRO J1655–40. MNRAS **427**, 595–606 (2012)

114. S. Motta, T. Muñoz-Darias, P. Casella, T. Belloni, J. Homan, Low-frequency oscillations in black holes: a spectral-timing approach to the case of GX 339–4. MNRAS **418**, 2292–2307 (2011)
115. S.E. Motta, T.M. Belloni, L. Stella, T. Muñoz-Darias, R. Fender, Precise mass and spin measurements for a stellar-mass black hole through X-ray timing: the case of GRO J1655–40. MNRAS **437**, 2554–2565 (2014)
116. S.E. Motta, P. Casella, M. Henze, T. Muñoz-Darias, A. Sanna, R. Fender, T. Belloni, Geometrical constraints on the origin of timing signals from black holes. MNRAS **447**, 2059–2072 (2015)
117. S.E. Motta, T. Muñoz-Darias, A. Sanna, R. Fender, T. Belloni, L. Stella, Black hole spin measurements through the relativistic precession model: XTE J1550-564, in *MNRAS* (2014)
118. T. Muñoz-Darias, M. Coriat, D.S. Plant, G. Ponti, R.P. Fender, R.J.H. Dunn, Inclination and relativistic effects in the outburst evolution of black hole transients. MNRAS **432**, 1330–1337 (2013)
119. T. Muñoz-Darias, R.P. Fender, S.E. Motta, T.M. Belloni, Black hole-like hysteresis and accretion states in neutron star low-mass X-ray binaries. MNRAS **443**, 3270–3283 (2014)
120. T. Muñoz-Darias, S. Motta, T.M. Belloni, Fast variability as a tracer of accretion regimes in black hole transients. MNRAS **410**, 679–684 (2011)
121. P. Mucciarelli, P. Casella, T. Belloni, L. Zampieri, P. Ranalli, A variable quasi-periodic oscillation in M82 X-1. Timing and spectral analysis of XMM-Newton and Rossi XTE observations. MNRAS **365**, 1123–1130 (2006)
122. J. Neilsen, J. Homan, A hybrid magnetically/thermally driven wind in the black hole GRO J1655–40? ApJ **750**, 27 (2012)
123. E. Nespoli, T. Belloni, J. Homan, J.M. Miller, W.H.G. Lewin, M. Méndez, M. van der Klis, A transient variable 6 Hz QPO from GX 339–4. Astron. Astrophys. **412**, 235–240 (2003)
124. M.A. Nowak, A. Juett, J. Homan, Y. Yao, J. Wilms, N.S. Schulz, C.R. Canizares, Disk-dominated states of 4U 1957+11: Chandra, XMM-Newton, and RXTE observations of ostensibly the most rapidly spinning Galactic black hole. ApJ **689**, 1199–1214 (2008)
125. J.F. Olive, D. Barret, L. Boirin, J.E. Grindlay, J.H. Swank, A.P. Smale, RXTE observation of the X-ray burster 1E 1724-3045. I. Timing study of the persistent X-ray emission with the PCA. Astron. Astrophys. **333**, 942–951 (1998)
126. T. Oosterbroek, M. van der Klis, J. van Paradijs, B. Vaughan, R. Rutledge, W.H.G. Lewin, Y. Tanaka, F. Nagase, T. Dotani, K. Mitsuda, S. Miyamoto, Spectral and timing behaviour of GS 2023+338. Astron. Astrophys. **321**, 776–790 (1997)
127. A.N. Parmar, T. Oosterbroek, L. Boirin, D. Lumb, Discovery of narrow X-ray absorption features from the dipping low-mass X-ray binary X 1624–490 with XMM-Newton. Astron. Astrophys. **386**, 910–915 (2002)
128. D.R. Pasham, T.E. Strohmayer, On the nature of the mHz X-ray quasi-periodic oscillations from ultraluminous X-ray source M82 X-1: search for timing-spectral correlations. ApJ **771**, 101 (2013)
129. J. Patterson, E.L. Robinson, R.E. Nather, Rapid and ultrarapid oscillations in RU Pegasi. ApJ **214**, 144–151 (1977)
130. F. Pintore, L. Zampieri, A. Wolter, T. Belloni, Ultraluminous X-ray sources: a deeper insight into their spectral evolution. MNRAS **439**, 3461–3475 (2014)
131. R.M. Plotkin, E. Gallo, P.G. Jonker, The X-ray spectral evolution of Galactic black hole X-ray binaries toward quiescence. ApJ **773**, 59 (2013)
132. R.M. Plotkin, E. Gallo, S. Markoff, J. Homan, P.G. Jonker, J.C.A. Miller-Jones, D.M. Russell, S. Drappeau, Constraints on relativistic jets in quiescent black hole X-ray binaries from broadband spectral modelling. MNRAS **446**, 4098–4111 (2015)
133. G. Ponti, R.P. Fender, M.C. Begelman, R.J.H. Dunn, J. Neilsen, M. Coriat, Ubiquitous equatorial accretion disc winds in black hole soft states. MNRAS **422**, L11 (2012)
134. G. Ponti, T. Muñoz-Darias, R.P. Fender, A connection between accretion state and Fe K absorption in an accreting neutron star: black hole-like soft-state winds? MNRAS **444**, 1829–1834 (2014)

135. J. Poutanen, J.H. Krolik, F. Ryde, The nature of spectral transitions in accreting black holes—The case of CYG X-1. MNRAS **292**, L21–L25 (1997)
136. D. Proga, T.R. Kallman, On the role of the ultraviolet and X-ray radiation in driving a disk wind in X-ray binaries. ApJ **565**, 455–470 (2002)
137. D. Psaltis, Probes and tests of strong-field gravity with observations in the electromagnetic spectrum. Living Rev. Relat. **11**, 9 (2008)
138. D. Psaltis, T. Belloni, M. van der Klis, Correlations in quasi-periodic oscillation and noise frequencies among neutron star and black hole x-ray binaries. APJ **520**, 262–270 (1999)
139. R.E. Pudritz, R. Ouyed, C. Fendt, A. Brandenburg, Disk winds, jets, and outflows: theoretical and computational foundations, in *Protostars and Planets V* (2007), pp. 277–294
140. P. Reig, T. Belloni, M. van der Klis, Does GRS 1915+105 exhibit "canonical" black-hole states? Astron. Astrophys. **412**, 229–233 (2003)
141. P. Reig, T. Belloni, M. van der Klis, M. Méndez, N.D. Kylafis, E.C. Ford, Phase lag variability associated with the 0.5–10 HZ quasi-periodic oscillations in GRS 1915+105. ApJ **541**, 883–888 (2000)
142. R.A. Remillard, M.P. Muno, J.E. McClintock, J.A. Orosz, Evidence for harmonic relationships in the high-frequency quasi-periodic oscillations of XTE J1550–564 and GRO J1655–40. ApJ **580**, 1030–1042 (2002)
143. R.A. Remillard, J.E. McClintock, X-ray properties of black-hole binaries. Ann. Rev. **44**, 49–92 (2006)
144. C.S. Reynolds, Constraints on compton-thick winds from black hole accretion disks: can we see the inner disk? ApJ **759**, L15 (2012)
145. L. Ruhlen, D.M. Smith, J.H. Swank, The nature and cause of spectral variability in LMC X-1. ApJ **742**, 75 (2011)
146. D.M. Russell, R.P. Fender, R.I. Hynes, C. Brocksopp, J. Homan, P.G. Jonker, M.M. Buxton, Global optical/infrared-X-ray correlations in X-ray binaries: quantifying disc and jet contributions. MNRAS **371**, 1334–1350 (2006)
147. D.M. Russell, J.C.A. Miller-Jones, T.J. Maccarone, Y.J. Yang, R.P. Fender, F. Lewis, Testing the jet quenching paradigm with an ultradeep observation of a steadily soft state black hole. ApJ **739**, L19 (2011)
148. T.D. Russell, J.C.A. Miller-Jones, P.A. Curran, R. Soria, D. Altamirano, S. Corbel, M. Coriat, A. Moin, D.M. Russell, G.R. Sivakoff, T.J. Slaven-Blair, T.M. Belloni, R.P. Fender, S. Heinz, P.G. Jonker, H.A. Krimm, E.G. Körding, D. Maitra, S. Markoff, M. Middleton, S. Migliari, R.A. Remillard, M.P. Rupen, C.L. Sarazin, A.J. Tetarenko, M.A.P. Torres, V. Tudose, A.K. Tzioumis, Radio monitoring of the hard state jets in the 2011 outburst of MAXI J1836–194. MNRAS **450**, 1745–1759 (2015)
149. M. Servillat, S.A. Farrell, D. Lin, O. Godet, D. Barret, N.A. Webb, X-ray variability and hardness of ESO 243–49 HLX-1: clear evidence for spectral state transitions. ApJ **743**, 6 (2011)
150. N.I. Shakura, R.A. Sunyaev, Black holes in binary systems. Observational appearance. Astron. Astrophys. **24**, 337–355 (1973)
151. M. Shidatsu, Y. Ueda, S. Nakahira, C. Done, K. Morihana, M. Sugizaki, T. Mihara, T. Hori, H. Negoro, N. Kawai, K. Yamaoka, K. Ebisawa, M. Matsuoka, M. Serino, T. Yoshikawa, T. Nagayama, N. Matsunaga, The accretion disk and ionized absorber of the 9.7 hr dipping black hole binary MAXI J1305-704. ApJ **779**, 26 (2013)
152. A.P. Smale, P.T. Boyd, Anomalous low states and long-term variability in the black hole binary LMC X-3. ApJ **756**, 146 (2012)
153. P. Soleri, T. Belloni, P. Casella, A transient low-frequency quasi-periodic oscillation from the black hole binary GRS 1915+105. MNRAS **383**, 1089–1102 (2008)
154. P. Soleri, R. Fender, On the nature of the 'radio-quiet' black hole binaries. MNRAS **413**, 2269–2280 (2011)
155. P. Soleri, T. Muñoz-Darias, S. Motta, T. Belloni, P. Casella, M. Méndez, D. Altamirano, M. Linares, R. Wijnands, R. Fender, M. van der Klis, A complex state transition from the black hole candidate Swift J1753.5-0127. MNRAS **429**, 1244–1257 (2013)

156. R. Soria, Hard and soft spectral states of ULXs. Astronomische Nachrichten **332**, 330 (2011)
157. R. Soria, J.W. Broderick, J. Hao, D.C. Hannikainen, M. Mehdipour, K. Pottschmidt, S.-N. Zhang, Accretion states of the Galactic microquasar GRS 1758–258. MNRAS **415**, 410–424 (2011)
158. L. Stella, M. Vietri, Lense-Thirring precession and quasi-periodic oscillations in low-mass X-ray binaries. ApJ **492**, L59 (1998)
159. L. Stella, M. Vietri, kHz quasiperiodic oscillations in low-mass X-ray binaries as probes of general relativity in the strong-field regime. Phys. Rev. Lett. **82**, 17–20 (1999)
160. L. Stella, M. Vietri, S.M. Morsink, Correlations in the quasi-periodic oscillation frequencies of low-mass X-ray binaries and the relativistic precession model. ApJ **524**, L63–L66 (1999)
161. H. Stiele, S. Motta, T. Muñoz-Darias, T.M. Belloni, Spectral Properties of Transitions Between Soft and Hard State in GX 339-4. ArXiv e-prints (2011)
162. A.M. Stirling, R.E. Spencer, C.J. de la Force, M.A. Garrett, R.P. Fender, R.N. Ogley, A relativistic jet from Cygnus X-1 in the low/hard X-ray state. MNRAS **327**, 1273–1278 (2001)
163. T.E. Strohmayer, Discovery of a 450 HZ quasi-periodic oscillation from the microquasar GRO J1655–40 with the Rossi X-Ray timing explorer. ApJ **552**, L49–L53 (2001)
164. T.E. Strohmayer, Discovery of a second high-frequency quasi-periodic oscillation from the microquasar GRS 1915+105. ApJ **554**, L169–L172 (2001)
165. T.E. Strohmayer, R.F. Mushotzky, Discovery of X-ray quasi-periodic oscillations from an ultraluminous X-ray source in M82: evidence against beaming. ApJ **586**, L61–L64 (2003)
166. M. Tagger, R. Pellat, An accretion-ejection instability in magnetized disks. Astron. Astrophys. **349**, 1003–1016 (1999)
167. M. Takizawa, T. Dotani, K. Mitsuda, E. Matsuba, M. Ogawa, T. Aoki, K. Asai, K. Ebisawa, K. Makishima, S. Miyamoto, S. Iga, B. Vaughan, R.E. Rutledge, W.H.G. Lewin, Spectral and temporal variability in the X-ray flux of GS 1124-683, Nova MUSCAE 1991. ApJ **489**, 272 (1997)
168. Y. Tanaka, W.H.G. Lewin, Black Hole Binaries (1995), pp. 126–174
169. H. Tananbaum, H. Gursky, E. Kellogg, R. Giacconi, C. Jones, Observation of a correlated X-ray transition in Cygnus X-1. ApJ **177**, L5 (1972)
170. L. Titarchuk, R. Fiorito, Spectral index and quasi-periodic oscillation frequency correlation in black hole sources: observational evidence of two phases and phase transition in black holes. ApJ **612**, 988–999 (2004)
171. Y. Ueda, H. Inoue, Y. Tanaka, K. Ebisawa, F. Nagase, T. Kotani, N. Gehrels, Detection of absorption-line features in the X-ray spectra of the Galactic superluminal source GRO J1655–40. ApJ **492**, 782–787 (1998)
172. M. van der Klis, Quasi-periodic oscillations and noise in low-mass X-ray binaries. Ann. Rev. **27**, 517–553 (1989)
173. M. van der Klis, Challenges in X-ray binary timing: current and future. Adv. Space Res. **34**, 2646–2656 (2004)
174. M. van der Klis, Overview of QPOs in neutron-star low-mass X-ray binaries. Adv. Space Res. **38**, 2675–2679 (2006)
175. P. Varnière, M. Tagger, Accretion-ejection instability in magnetized disks: feeding the corona with Alfvén waves. Astron. Astrophys. **394**, 329–338 (2002)
176. P. Varnière, M. Tagger, J. Rodriguez, A possible interpretation for the apparent differences in LFQPO types in microquasars. Astron. Astrophys. **545**, A40 (2012)
177. N. Webb, D. Cseh, E. Lenc, O. Godet, D. Barret, S. Corbel, S. Farrell, R. Fender, N. Gehrels, I. Heywood, Radio detections during two state transitions of the intermediate-mass black hole HLX-1. Science **337**, 554 (2012)
178. R. Wijnands, J. Homan, M. van der Klis, The complex phase-lag behavior of the 3–12 HZ quasi-periodic oscillations during the very high state of XTE J1550–564. ApJ **526**, L33–L36 (1999)
179. R. Wijnands, M. van der Klis, The broadband power spectra of X-ray binaries. ApJ **514**, 939–944 (1999)

180. J. Wilms, K. Pottschmidt, G.G. Pooley, S. Markoff, M.A. Nowak, I. Kreykenbohm, R.E. Rothschild, Correlated radio-X-ray variability of Galactic black holes: a radio-X-ray flare in Cygnus X-1. ApJ **663**, L97–L100 (2007)
181. Q. Wu, M. Gu, The X-ray spectral evolution in X-ray binaries and its application to constrain the black hole mass of ultraluminous X-ray sources. ApJ **682**, 212–217 (2008)
182. Y.X. Wu, W. Yu, Z. Yan, L. Sun, T.P. Li, On the relation of hard X-ray peak flux and outburst waiting time in the black hole transient GX 339–4. Astron. Astrophys. **512**, A32 (2010)
183. A.A. Zdziarski, P. Lubiński, M. Gilfanov, M. Revnivtsev, Correlations between X-ray and radio spectral properties of accreting black holes. MNRAS **342**, 355–372 (2003)

Chapter 3
Black Hole Spin: Theory and Observation

M. Middleton

Abstract In the standard paradigm, astrophysical black holes can be described solely by their mass and angular momentum—commonly referred to as 'spin'—resulting from the process of their birth and subsequent growth via accretion. Whilst the mass has a standard Newtonian interpretation, the spin does not, with the effect of nonzero spin leaving an indelible imprint on the space-time closest to the black hole. As a consequence of relativistic frame-dragging, particle orbits are affected both in terms of stability and precession, which impacts on the emission characteristics of accreting black holes both stellar mass in black hole binaries (BHBs) and supermassive in active galactic nuclei (AGN). Over the last 30 years, techniques have been developed that take into account these changes to estimate the spin which can then be used to understand the birth and growth of black holes and potentially the powering of astrophysical jets. In this chapter, we provide a broad overview of both the theoretical effects of spin, the means by which it can be estimated and the results of ongoing campaigns.

3.1 Preface

In the generally accepted model of Einstein's General Relativity (GR—although modified forms of GR cannot yet be ruled out), black holes (BHs) are defined by only their mass, charge and angular momentum, hereafter referred to as spin. In an astrophysical setting, the charge will soon neutralise (or else be radiated away during the formation process in accordance with Price's theorem) and so BHs can be entirely defined by only their mass and spin—this is the often-touted 'no-hair' theorem. Whilst the effect of mass is relatively benign from the standpoint of observation and theory (generally acting only to scale the energetics, e.g. Shakura and Sunyaev [301]—and accretion-related timescales—e.g. McHardy et al. [195]), the spin has much more to offer in terms of revealing how BHs formed and grew, how accretion

M. Middleton (✉)
Institute of Astronomy, Madingley Road, Cambridge, UK
e-mail: mjm@ast.cam.ac.uk

© Springer-Verlag Berlin Heidelberg 2016   99
C. Bambi (ed.), *Astrophysics of Black Holes*, Astrophysics
and Space Science Library 440, DOI 10.1007/978-3-662-52859-4_3

operates in the regime of strong gravity and how the most powerful ejections of material in the universe may be powered.

What follows should not be viewed as exhaustive but a summary of the state of the art. In Sect. 3.2, we will review the necessary theory to make sense of the observations and methods that have been applied in attempts to measure BH spin in various systems under various assumptions. In Sect. 3.3, we will discuss the traditional techniques used to measure the spin in BHBs and AGN and the implications of these measurements (with special attention paid to the launching of astrophysical jets). Sections 3.4 and 3.5 discuss new techniques which incorporate the time domain, and in Sect. 3.6, we conclude with some final remarks about the future prospects for the field.

3.2 Useful Theory

Before we discuss how methods to estimate the spin have developed and been applied to observation, it is important to understand how a spinning BH affects the space-time in which it is embedded as the outcome dictates our approach to studying BHs. Below, we present the formalisms which govern the nature of space-time around such a BH and the behaviour of test particles in close proximity (at large distances, this tends towards a Newtonian description); we stress that a deep working knowledge of GR is not necessary to appreciate what follows, with only those formulae considered relevant for our later discussions being presented (though for the more experienced reader, we suggest the review of Abramowicz and Fragile [1]).

The solution to Einstein's field equations for a spherically symmetric, nonrotating massive body was discovered by Karl Schwarzschild in 1916 for which the metric, which describes the geometry of empty space-time (a 'manifold'), is named, whilst the generalisation to a rotating (uncharged) BH is known as the Kerr metric after Roy Kerr, who discovered the solution in 1963 (see his explanation in Kerr [150]). The difference between the two solutions results from the inclusion of the BH angular momentum ($\mathbf{J}$) which, as we will see, has a significant impact on the orbit of test particles and the behaviour of infalling material. The metric is usually presented in Boyer–Lindquist coordinates (t, r, θ, ϕ) which can be interpreted as spherical polar coordinates and is related to cartesian coordinates via the following standard transforms:

$$x = \sqrt{r^2 + a^2} \sin \theta \cos \phi$$
$$y = \sqrt{r^2 + a^2} \sin \theta \sin \phi$$
$$z = r \cos \theta$$

Here, a is the BH-specific angular momentum ($a = |J|/M$) although in literature discussing observation, is often expressed in its dimensionless form, a_*:

$$a_* = \frac{|J|c}{GM^2} \tag{3.1}$$

where M is the BH mass in solar units and G and c are the usual constants.

In flat (Minkowski) space-time, a line element of length ds is simply given by $ds^2 = -(cdt)^2 + dx^2 + dy^2 + dz^2$; however, in the Kerr space-time (in natural units of $G = c = 1$), this becomes:

$$ds^2 = -\left(1 - \frac{2Mr}{\rho^2}\right)dt^2 + \frac{\rho^2}{\Delta}dr^2 + \rho^2 d\theta^2 + \left(r^2 + a^2 + \frac{2Mra^2}{\rho^2}\sin^2\theta\right)\sin^2\theta d\phi^2$$
$$- \frac{4Mra}{\rho^2}\sin^2\theta d\phi dt \tag{3.2}$$

where

$$\rho^2 = r^2 + a^2 \cos^2\theta \tag{3.3}$$

$$\Delta = r^2 - 2Mr + a^2 \tag{3.4}$$

Using Einstein's notation, this can be written in terms of the covariant metric tensor, $g_{\mu\nu}$, where $ds^2 = g_{\mu\nu}dx^\mu dx^\nu$ and $g_{\mu\nu}$ in vector form is:

$$\begin{pmatrix} g_{tt} & 0 & 0 & g_{t\phi} \\ 0 & g_{rr} & 0 & 0 \\ 0 & 0 & g_{\theta\theta} & 0 \\ g_{t\phi} & 0 & 0 & g_{\phi\phi} \end{pmatrix}$$

By inspection, the components of the metric in Eq. 3.2 are then:

$$g_{tt} = -\left(1 - \frac{2Mr}{\rho^2}\right) \tag{3.5}$$

$$g_{t\phi} = \frac{-4Mr}{\rho^2}a\sin^2\theta \tag{3.6}$$

$$g_{\phi\phi} = \left(r^2 + a^2 + \frac{2Mra^2}{\rho^2}\sin^2\theta\right)\sin^2\theta \tag{3.7}$$

$$g_{rr} = \frac{\rho^2}{\Delta} \tag{3.8}$$

$$g_{\theta\theta} = \rho^2 \tag{3.9}$$

and setting $a = 0$ returns the metric for a nonspinning, Schwarzschild BH.

Whist the above may seem mathematically daunting, the components of the metric play a vital role in allowing us to predict the effect of nonzero spin. In the Schwarzschild metric, there are two important radii to consider: the first is the position of

the static surface (also called a 'null hypersurface') of the event horizon, where an
observer in a distant reference frame would observe a body, travelling at the speed of
light radially away from the BH to be stationary. In this case (i.e. for a nonspinning
BH), the event horizon is located at the Schwarzschild radius (which can be easily
derived in the Newtonian case with the escape velocity set equal to the speed of light),
$R_s = 2GM/c^2$ (=$2R_g$ or $2M$ in natural units). The second important radius we need
to consider is the position of the marginally stable orbit more commonly referred to
as the innermost stable circular orbit (ISCO), inside of which stable orbits are not
possible and any accreted material takes a laminar plunge to the event horizon on a
dynamical timescale. The position of the ISCO can be found from considering the
radial equation of geodesic motion using the proper time (i.e. that measured by a
clock at rest) τ:

$$\frac{d^2r}{d\tau^2} + \frac{M}{r^2} - (1 - 3M/r)\, u_\phi^2/r^3 = 0 \qquad (3.10)$$

where u_ϕ is the specific angular momentum. If we consider circular orbits, $d^2r/d\tau^2$
$= 0$ so $u_\phi^2 = Mr^2/(r - 3M)$, the specific angular momentum has a minimum at r =
$6M$. Physically, this is distinct from the Newtonian case where angular momentum
decreases monotonically down to the central object. At radii within the ISCO, circular
orbits are no longer possible (see the discussion in Abramowicz and Fragile [1]) and
the material undergoes radial free fall to the event horizon and the singularity beyond.

Nonzero values of the spin indicate a Kerr BH, with a positive value corresponding to the BH rotating in the same direction as the orbiting particles around it, i.e.
prograde. Conversely, negative spin values indicate that the orbits are oriented in
the opposite direction, i.e. retrograde to the BH spin. Based on the third law of BH
thermodynamics which states that a BH cannot have zero surface gravity [17], the
spin must have natural limiting values of -1 and 1 (at which point the surface gravity
is zero). An additional constraint arises from the consideration of the Kerr solution
and by setting Eq. 3.4 equal to zero: $\Delta = r^2 - 2Mr + a^2 = 0$, which is a coordinate
singularity in Eq. 3.2. It can be readily seen that there are no real solutions when
$a^2 > M^2$, which implies that in such a case, there is no horizon and no BH, leading
to a 'naked singularity' which is forbidden (due to paradoxes); so once again, we
find a limiting value of $|a| < M$ or (as $a_* = a/M$), $|a_*| < 1$ as before.

Assuming that, at the point of BH formation, the spin is less than maximal (in the
case of stellar mass BHs, this is eminently sensible, given that angular momentum
may be lost in the supernova explosion), accretion of matter in the prograde direction
will both increase the mass and 'spin-up' the BH, analogous to the situation of
accreting neutron stars (e.g. Bisnovatyi-Kogan and Komberg [26]). Considering only
the accretion of matter, the BH's growth follows Bardeen's law [17]; however, as
discussed in Thorne [325], the effect of radiation is important. If one were to ignore
the radiation from the disc, then in principle the spin could reach a limiting value
of 1, however, as pointed out by Bardeen et al. [17], the capture cross section for
photons with oppositely aligned momentum is larger than when aligned and so these
photons will act to buffer against the spin reaching unity. More accurately, above

$a_* = 0.90$, radiation effects *cannot* be ignored and lead to a deviation in evolution away from Bardeen's law such that the limiting, 'maximal' value is reached at a_* = 0.998 for prograde spin or -0.998 in the retrograde case (and changes only very little depending on the nature of the illumination: [325]). It is important to note that this maximal value does not account for the effect of torques which are expected to result from magnetic fields threading the plunging region which may act to reduce the maximum spin that can be achieved [97].

It is important to note that whilst accretion must inevitably change the spin (unless maximal already), we cannot yet observe this on any human timescale. From Bardeen's law, it can be seen that it would take more than the mass of the BH itself to be accreted to change the spin from 0 to 1; given typical mass transfer rates in BHBs via Roche Lobe overflow of 10^{-6}–$10^{-7} M_\odot$/year and outburst duty cycles of $\sim 1\%$ (e.g. Fragos et al. [92]), it is clear that we will have to wait around the lifetime of the binary itself ($\sim$ billions of years) to see such a substantial change in the spin of stellar mass BHs (and equally for supermassive BHs—SMBHs—in AGN where the duty cycle is thought to be similar, but mass loss via winds could be substantial). Smaller changes are, however, possible on smaller fractions of the binary's lifetime, depending on the starting mass and spin of the black hole (see Fragos and McClintock [93]); however, for changes in the spin that occur on observable timescales, we presently—and for the foreseeable future—will lack the ability to detect them via the methods we will discuss in the forthcoming sections of this chapter.

3.2.1 Frame-Dragging

A key consequence of having a spinning BH—in terms of observational implications—is the concept of relativistic frame-dragging. As a result of the BH's nonzero angular momentum, space-time moves (is frame-dragged) in the direction of the spin in its vicinity, thereby imparting energy to an orbiting test particle. This can be seen directly when we consider an observer, with (contravariant) four velocity u^μ ($= dx^\mu/d\tau$) who falls into the BH with zero angular momentum or $L = u_\phi = 0$. This is the usual definition of a zero angular momentum observer (ZAMO). The contravariant component of the velocity is nonzero (except as $r \to 0$), so $u^\phi = g^{\phi t} u_t \neq 0$. The angular velocity of the ZAMO is as follows:

$$\Omega = \frac{d\phi}{dt} = \frac{\frac{d\phi}{d\tau}}{\frac{dt}{d\tau}} = \frac{u^\phi}{u^t} \neq 0 \tag{3.11}$$

Ω can be computed from $u_\phi = 0 = g_{\phi\phi} u^\phi + g_{\phi t} u^t$ which gives:

$$\Omega = \frac{u^\phi}{u^t} = -\frac{g_{\phi t}}{g_{\phi\phi}} \tag{3.12}$$

Substituting Eqs. 3.5 and 3.6 for the components of the metric leads to:

$$\Omega = \frac{2Mar}{(r^2 + a^2)^2 - a^2 \Delta \sin^2 \theta} \tag{3.13}$$

By substituting Eq. 3.4, we can see that $(r^2 + a^2)^2 > a^2(r^2 - 2Mr + a^2) \sin^2 \theta$ and so $\Omega/Ma > 0$. Therefore, as a result of the nonzero BH spin, a ZAMO is forced to co-rotate (frame-dragged) in the direction of its rotation (as the angular velocity has the same sign as the angular momentum).

As a consequence of frame-dragging, whilst the Schwarzschild metric is only singular (in Boyer–Lindquist coordinates) at the static surface of the event horizon, solutions are singular across two null hypersurfaces in the Kerr metric. The new position of the event horizon is found where g_{rr}, the radial component of the metric (Eq. 3.2) tends to infinity. By setting $1/g_{rr} = 0$, we can see that this is the same as the solution to the coordinate singularity at $\Delta = 0 = r^2 - 2Mr + a^2$. Solving the quadratic leads to the solution for the radius of the horizon:

$$r_H = M \pm \sqrt{M^2 - a^2} \tag{3.14}$$

The positive solution of this equation defines the event horizon (as radii below this are forced to travel faster than the speed of light). As we can see, for nonzero spin, the position of the event horizon can be within the Schwarzschild radius. The second hypersurface occurs when g_{tt} changes sign or, from Eqs. 3.4 and 3.5:

$$g_{tt} = 0 = - \left(1 - \frac{2Mr}{r^2 + a^2 \cos^2 \theta} \right) \rightarrow 0 = r^2 - 2Mr + a^2 \cos^2 \theta \tag{3.15}$$

which has the solutions:

$$r_E = M \pm \sqrt{M^2 - a^2 \cos^2 \theta} \tag{3.16}$$

The surface at r_{E+} is referred to as the 'ergosphere' and the region between this and the event horizon, the ergoregion. It is easy to see that $\sqrt{M^2 - a^2 \cos^2 \theta} > \sqrt{M^2 - a^2}$ for all co-latitudes (θ) except at the poles where the position of the event horizon and ergosphere meets (see Fig. 3.1).

As $r_{E+} > r_{H+}$ (when $\theta \neq 0$ and $\theta \neq \pi$), an observer within the ergosphere can still be in causal contact with the outside universe; this is plotted in Fig. 3.2. Within the ergosphere, it is not possible for a physical observer to remain at rest, and from calculating the effect of orbits within this region, it was discovered that negative energy (retrograde) trajectories/orbits are possible (see, e.g. Penrose [258]; Bardeen and Press [18]). Should an orbiting body fragment within the ergosphere, then the total energy of those fragments not induced into negative energy orbits will be greater, having effectively tapped the energy (angular momentum) of the BH. This tapping of the BH's spin is called the Penrose effect [258] and the magnetic field analog,

Fig. 3.1 2D positions of the event horizon (Eq. 3.14: *dashed line*) and ergosphere (Eq. 3.16: *solid line*) as a function of BH spin (across polar coordinate space—where the x-axis is the radial distance from the BH). The hypersurfaces meet at the poles (co-latitude of 0 degrees) with the ergosphere 'pinched' down as the spin increases

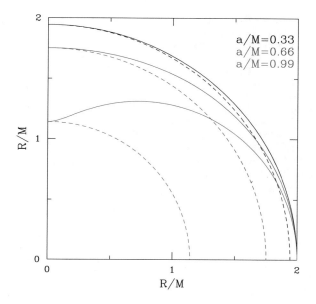

the Blandford–Znajek effect [29], which we shall discuss in Sect. 3.3.5 in relation to powering 'superluminal' ejections.

As stable orbits are possible closer to the BH for prograde spin [18], frame-dragging changes the position of the ISCO which is a well-defined, monotonic function of a/M [18] and is shown in Fig. 3.2:

$$r_{isco} = M \left[3 + Z_2 \mp \left[(3 - Z_1)(3 + Z_1 + 2Z_2) \right]^{1/2} \right] \tag{3.17}$$

where

$$Z_1 = 1 + \left(1 - a^2/M^2 \right)^{1/3} \left[(1 + a/M)^{1/3} + (1 - a/M)^{1/3} \right] \tag{3.18}$$

$$Z_2 = \left(3a^2/M^2 + Z_1^2 \right)^{1/2} \tag{3.19}$$

where $\mp$ is used the top sign refers to a treatment where the spin is prograde and the bottom sign to where the spin is retrograde.

As is shown in Fig. 3.2, for $a > 0$, the position of the ISCO lies within that for a Schwarzschild BH ($6M$), reaching a minimum at $1.235M$ (for $a_* = 0.998$), and is pushed further out in the case of $a < 0$ (retrograde spin) towards a maximum of $9M$. As we shall see in the following sections, this correspondence between the spin and location of the ISCO allows us to construct models to estimate the spin from observation.

Fig. 3.2 Position of the
respective hypersurfaces in
the Kerr metric as a function
of BH spin (Eqs. 3.14, 3.16
and 3.17), R_H, the event
horizon, R_E, the ergosphere
(shown for $\theta = \pi/2$, i.e. in
the plane of an accretion
disc) and R_{isco}, the innermost
stable circular orbit. The *top
right* inset shows an enlarged
version covering the highest
spins where R_{isco} sits *inside*
the ergosphere

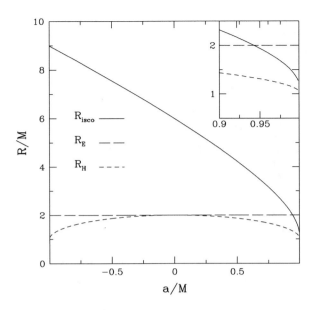

It is worth noting as a point of general interest that in addition to the changes
to the positions of the hypersurfaces discussed above, the singularity itself can no
longer be point-like but must take the form of a ring (we will not discuss the effects
of ring singularities further but point the interested reader to Burko and Ori [36]).
Additionally, although we have discussed the effect of frame-dragging on test particles, electromagnetic fields (generated either in the flow via the Magnetorotational
instability (MRI): [14] or in the local environment to the BH) will also be affected
and can lead to reconnection events and particle acceleration (e.g. Karas et al. [148]).

3.2.1.1 Relativistic Precession

There are two further implications of a spinning BH, resulting from the effect of
frame-dragging: Lense–Thirring precession and the Bardeen–Peterson effect, both
of which may provide important observational diagnostics of the spin and region of
strong gravity close to the BH.

Lense–Thirring precession (also called the Lense–Thirring effect: [169]) describes
the behaviour of orbiting and vertically displaced motion in proximity to a rotating
massive body. For this reason, it is relevant not only for Kerr BHs but more generally
for satellites of astrophysical bodies such as stars and planets. Due to frame-dragging,
the orbital motion undergoes precession leading to epicyclic oscillations about peri-
apsis (position of closest approach) and the ecliptic as long as the orbit is vertically
misaligned with the rotation axis of the rotating body.

Should the accretion disc be misaligned with the BH spin axis (for instance due to a supernova kick, see Brandt and Podsiadlowski [31]), then the precession of orbits due to the Lense–Thirring effect produces a torque. If this torque is larger than the viscous torques in the disc, a fluid (nonsolid body) inner disc will align perpendicular to the BH spin axis, whilst beyond the warp radius the disc aligns with the binary orbit; this is the Bardeen–Petterson effect [19]. Should the torques not be dissipated (though see discussions by Armitage and Natarajan [10]; Marković and Lamb [185]), the whole of the inner disc can precess leading to important effects on the emergent spectrum and jets (see Sect. 3.4). Such precession may lead to clear signatures in the time domain as well as the energy domain [130], and as we will discuss later, can provide a diagnostic for the BH spin.

The manner in which the disc warp created by the Bardeen–Peterson effect propagates depends on the nature of the disc: if it is thin or the viscosity high, then the warp diffuses (due to viscous torques as discussed by Pringle [261]), whilst if the disc is thick or viscosity low, the warp propagates as a wave [253]. King et al. [151] discuss the general case (for both viscosity cases and for the full range of disc tilts and misalignment) of how the torque between the BH and disc as a result of precession can lead to co-alignment or counteralignment of the BH/disc system. The authors conclude that alignment depends on the detailed properties of the disc, namely how the warp is propagated, although, in general, on short timescales, it is possible that the disc tries to misalign with the hole (essentially spinning the hole down), whilst on long timescales there is a tendency towards co-alignment (and spin-up). Making comparisons to observation, Maccarone [177] reports that the jet angle in the BHBs, GRO J1655-40 and SAX J 1819–2525, is misaligned with respect to (i.e. not perpendicular to) the binary orbit. Assuming the jet angle is tied to the inner disc and the spin of the BH, this may indicate that the misalignment resulted from the formation process (e.g. Brandt and Podsiadlowski [31]) and the time required for alignment is potentially a significant fraction of the binary lifetime [187, 309]. King et al. [151] note that in such systems, the angular momentum of the BH is much larger than that of the disc and so the crucial timescale is that on which tidal forces can transfer angular momentum from the binary orbit to the disc—on timescales shorter than this, counteraligned discs may be possible.

In the case of AGN, the picture is less clear as the timescales for accretion-driven changes (and alignment) are considerably longer (typically scaling with mass), whilst the means by which material reaches the inner sub-pc disc is still debated. Should material fall through the galactic disc, then the angular momentum is expected to be in a single direction and the BH's spin axis should appear aligned with the host galaxy's stellar disc (assuming that the growth is driven by accretion rather than via BH–BH mergers). Instead, should material with a range of angular momenta be accreted (e.g. via condensed filaments: [236])—a situation often referred to as 'chaotic' accretion—then it is possible that the disc-BH system will be initially misaligned with respect to the inflowing material (which is assumed to be misaligned with the host galaxy's stellar disc) and then co-align—as a consequence, the sub-pc system does not need to be aligned with the galactic plane. Observations of jets (see Hopkins et al. [125]) and the inclination of the sub-pc disc determined from reflection spectroscopy (Middleton et al. [202]), indicates that many AGN-galaxy systems do indeed appear misaligned which likely points towards a recent chaotic accretion history.

3.3 Observational Tests of Spin I—the Energy Spectral Domain

The effect of frame-dragging and the change in the position of the ISCO with spin (Fig. 3.2) have led to the development of methods by which the BH spin can be *estimated*. We purposely use the word *estimated* here to signify the uncertainty inherent in characterising such an intrinsic yet complex property, relying on models which are themselves based on caveats and assumptions. However, this should *not* be read as a criticism of efforts both past and ongoing to better estimate and constrain the spin, rather that the reliability of a chosen method should be evaluated against the backdrop of systematic uncertainties.

3.3.1 Modelling the Continuum (Disc) Spectrum

As mentioned elsewhere in this compilation, transient (predominantly low-mass companion) BHBs undergo outburst cycles regulated by disc instabilities (e.g. Lasota [168]) evolving in brightness and spectral shape (see McClintock and Remillard [192] and Done et al. [69] for reviews). Towards the peak of the outburst, the spectrum becomes increasingly dominated by emission originating from the accretion disc. Under the assumption that the inner disc radius (R_{in}) sits at the ISCO (as seen in GRMHD simulations: [257, 298]—and see Zhu et al. [345] for the effect of the plunging region—and evidenced in the observational study of LMC X-3 by Steiner et al. [307]), we can show that the properties (namely the temperature and luminosity) are related to the spin through the following formulae (for a rigorous discussion, we point the reader to Frank et al. [94]). We point out that the following derivations are meant only to illustrate the case using the simplest, nonrelativistic treatment (we discuss the relativistic disc modelling in Sect. 3.3.1.1). The presence of viscous torques on the differential (Keplerian) orbits leads to dissipation of mechanical energy with the torque defined as:

$$t_\phi(R) = 2\pi R \nu \Sigma R^2 \qquad (3.20)$$

where ν is the kinematic viscosity, Σ is the surface density and Ω' is the radial gradient of angular momentum ($d\Omega/dr$).

Although ν becomes unimportant in terms of calculating the emission profile (as we shall soon see), its form is relatively important historically as it can also be parameterised as:

$$v = \alpha c_s H \tag{3.21}$$

where c_s is the sound speed, H is the height of the disc from the mid-plane and α is the viscosity parameter that underpins the α-prescription of Shakura and Sunyaev [301]. This formula assumes that turbulence drives the viscosity and results from a consideration of the typical size of a turbulent eddy (which must be less than the disc scale height, H/R) and the assumption that the turbulent velocity is not supersonic. As a consequence, we would expect $\alpha < 1$. As Frank et al. [94] point out, this is not a physical statement as the true nature of the viscosity is unknown (although they also point out that magnetic stress would also lead to $\alpha < 1$).

Irrespective of the nature of the viscosity, the amount of mechanical heat loss is given by $t_\phi(R)\Omega' dR$ and is dissipated across both sides of the disc ($2 \times 2\pi R dR$), giving a heat loss per unit area ($D(R)$) of:

$$D(R) = \frac{t_\phi \Omega'}{4\pi R} = \frac{v\Sigma}{2}(R\Omega')^2 \tag{3.22}$$

Setting Ω to be Keplerian (i.e. differential rotation: $\Omega_k = (GM/R^3)^{1/2}$), gives:

$$D(R) = \frac{9}{8} v\Sigma \frac{GM}{R^3} \tag{3.23}$$

From conservation of mass and angular momentum, and assuming zero torque at the innermost edge of the disc (i.e. $\Omega' = 0$ at $R = R_{in}$; see Krolik [161]; Gammie [96] and Balbus [13] for issues associated with this assumption), it can be seen that:

$$v\Sigma = \frac{\dot{M}}{3\pi}\left[1 - \left(\frac{R_{in}}{R}\right)^{1/2}\right] \tag{3.24}$$

Combining Eqs. 3.23 and 3.24 leads to the formula for the dissipation of energy per unit area more commonly seen in the literature:

$$D(R) = \frac{3GM\dot{M}}{8\pi R^3}\left[1 - \left(\frac{R_{in}}{R}\right)^{1/2}\right] \tag{3.25}$$

which importantly demonstrates that the heating is *independent* of viscosity (either related to the sum of radiation and gas pressure in the disc: [301], or via magnetic stresses caused by the MRI: [14]). Assuming the disc to be fully optically thick to the emergent thermal radiation, we should expect local emission (i.e. at each radius) to be a blackbody (Planck's distribution) with a peak, effective temperature, T_{eff} according to Stefan–Boltzmann's law:

$$D(R) = \sigma_{SB} T_{eff}^4 \tag{3.26}$$

or

$$T_{eff} = \left\{ \frac{3GM\dot{M}}{8\pi R^3 \sigma_{SB}} \left[1 - \left(\frac{R_{in}}{R} \right)^{1/2} \right] \right\}^{1/4} \tag{3.27}$$

Thus, in this nonrelativistic approximation, the temperature of the thermal emission is in principle related to the position of the inner radius and is therefore a diagnostic of the spin.

The luminosity that emerges as a result of the process of accretion through the disc can be approximated by:

$$L = \frac{GM\dot{M}}{2R} \tag{3.28}$$

where R is the position of the inner edge of the flow and the factor of 2 in the denominator results from the virialisation of the system (i.e. half of the potential energy is radiated, whilst the remaining half is converted into kinetic energy and lost to the BH). This luminosity can also be parameterised as the conversion of rest mass to energy given by:

$$L = \eta \dot{M} c^2 \tag{3.29}$$

where η is the radiative efficiency. By equating the two formulae above, it is clear that in the simplest picture, the radiative efficiency is a function of the position of the inner edge and therefore the spin, ranging from $\sim$8–40 % for zero through to maximal (prograde) spin. This is the simplified Newtonian case and is a good approximation to the actual efficiency as a function of spin which goes as:

$$\eta = 1 - (R_{ISCO} - 2M \pm A_1)(R_{ISCO} - 3M \pm 2A_1)^{-1/2} \tag{3.30}$$

where $A_1 = a\sqrt{M/R_{ISCO}}$. We note that the above equations assume that the mass falling onto the BH is only converted into radiation. This is demonstrably not the case as powerful winds and jets are ubiquitous to accretion flows; however, this remains an important illustrative point and a useful theoretical framework for discussing BH accretion discs.

In practice, the emission spectrum from the accretion disc is a convolution of the thermal emission from all radii or a 'multicolour' disc blackbody (e.g. Mitsuda et al. [223]; Makishima et al. [181]). In addition to this deviation from a blackbody, a further complication arises due to the effect of opacity which determines how deep into the disc atmosphere we observe, i.e. the position of the 'effective photosphere', τ_{eff}. The two 'competing' forms of opacity are electron scattering (κ_T) and absorption via both free-free (κ_{ff}) and via metal edges/bound-free transitions (κ_{bf}). Whilst κ_T is independent of temperature and density, both forms of absorption opacity scale as $\rho * T^{-7/2}$ (Kramer's law). Thus, the position of $\tau_{eff} \approx \tau_T \sqrt{\kappa_{abs}/\kappa_T} = 1$ is a function of temperature/frequency (where κ_{abs} is the sum of the contributions to the absorption opacity). At higher frequencies, we can see further into the disc as the absorption opacity is lower; as there is a negative, vertical temperature gradient through the disc, when $\tau_{eff} = 1$ is further into the disc, we observe a larger offset in temperature compared to the surface. Such effects lead to the requirement of a colour correction/spectral hardening factor, f_{col} where the *observed* temperature of a blackbody at a given radius is $T_{col} = f_{col} * T_{eff}$ and the intensity at a given frequency (I_ν):

$$I_\nu = \frac{1}{f_{col}^4} B_\nu(T_{col}) \tag{3.31}$$

where B_ν is the Planck function and f_{col} can be roughly parameterised by the ratio (i.e. relative importance) of the competing opacities in the disc:

$$f_{col} \sim \left(\frac{\kappa_{tot}}{\kappa_{abs}}\right)^{1/4} \tag{3.32}$$

where $\kappa_{tot} = \kappa_{abs} + \kappa_T$ and f_{col} reaches saturation [56] at:

$$f_{col} \sim \left(72 \, keV / T_{eff}\right)^{1/9} \tag{3.33}$$

The simplest and most widely adopted disc model, DISKBB (for use in the spectral fitting package XSPEC: [12] or ISIS: [126]), takes a value of $f_{col} = 1.7$ [304], assuming electron scattering dominates the opacity (see, e.g. Ebisuzaki et al. [75]). Although density and temperature (giving the relative balance of opacities) and therefore f_{col} are unlikely to be constant across the disc (e.g. Gierliński and Done [110]), this model is commonly used to describe the thermal emission seen in BHBs and neutron star binaries and can be used to provide a crude estimate of R_{in} and therefore the BH spin.

3.3.1.1 Beyond the Simple Picture

A more accurate picture of accretion in the framework of GR was developed by Novikov and Thorne [246] and Page and Thorne [252], assuming a razor-thin disc and zero-torque inner boundary condition (and can be seen as the relativistic analog to the Shakura–Sunyaev disc), more commonly referred to as the general relativistic accretion disc (GRAD) model. Building upon this relativistic framework, models are now available that include the full 'suite' of relativistic corrections (Doppler boosting and gravitational redshift), the effect of returning radiation and importantly nonzero inner boundary conditions (i.e. $t_\phi(R_{in}) \neq 0$). This last point is hotly debated as magnetic fields crossing the ISCO may connect the disc to the BH or plunging region and thereby provide a torque (see discussions by Paczyński [250]; Armitage et al. [11]; Hawley and Krolik [119]; Afshordi and Paczyński [3]; Li [170]). One of the most widely used of the GRAD models is KERRBB [171] which includes a grid of spectra created via 'ray-tracing' in the Kerr metric. The method of ray-tracing is a well-established and reliable means of mapping photon paths in a given metric and can be seen as a way to effectively visualise emission from the accretion flow (e.g. Cunningham and Bardeen [47]; Cunningham [48]; Rauch and Blandford [266]; Fanton et al. [86]; Čadež et al. [38]; Müller and Camenzind [232]; Schnittman and Bertschinger [296]; Dexter and Agol [62]). In practice, the disc 'image' seen by an observer in some observer-system geometry is broken into a number of small elements and photon paths are traced back to the disc. By assuming a local flux density profile at each location in the disc and by incorporating relativistic effects (Doppler boosting and gravitational redshift), the final spectrum can be reconstructed by summing over the disc elements the paths intercept. In addition to direct illumination, ray-tracing can also track photons which return to the disc from the far side due to gravitational light-bending, leading to a change in the locally emitted flux. The grids in the KERRBB model allow for a range of spin, inclination and BH mass whilst assuming a standard disc structure with constant f_{col} (= 1.7, although this value can be changed in the model) and allows for limb-darkening (e.g. Svoboda et al. [319]).

In determining f_{col} in the above models, bound-free absorption has been ignored; however, it can dominate over free-free opacity and lead to changes in f_{col} with an increased likelihood that photons are instead 'destroyed' rather than propagated and scattered. At the time of writing, the only disc model which incorporates metal edges is BHSPEC [57, 58]. This model describes a GRAD [246] but *unlike* KERRBB calculates the disc spectrum by including a relativistic transfer function in place of ray-tracing. The transfer function provides the integration kernel in calculating the disc emission and contains information regarding the Doppler boost due to rotation and gravitational light-bending (see Cunningham [48, 49]; Laor [167]; Speith et al. [306]; Agol [4]; Agol and Krolik [5]; Dovčiak et al. [71]).

Davis et al. [57] create the BHSPEC model tables following the methods laid out in Hubeny et al. [127, 128] (and references therein) adopting the TLUSTY stellar atmosphere code [129] to solve the equations for the vertical structure and angular dependence of the radiative transfer. In addition to including bound-free opacities (assuming ground state populations), the model fully accounts for Comptonisation of the escaping radiation. The authors find that bound-free absorption does indeed play an important role in determining f_{col} with typical values between 1.5 and 1.6 (lower than found by Shimura and Takahara [304] and Merloni et al. [199]), with the actual value depending on the mass accretion rate and spin (and weakly on the adopted value of the α parameter or stress prescription). As with KERRBB, BHSPEC accounts for the angular dependence of limb-darkening.

The hybrid code, KERRBB2, combines the ability of BHSPEC to self-consistently determine spectral hardening values with the ability of KERRBB to account for returning radiation. To do this, tables of spectral hardening values are precomputed (publicly released for *RXTE* only) by fitting spectra generated with BHSPEC with KERRBB at fixed sets of input values (see McClintock and Remillard [192] for an example of its use).

In the case of the above models, it is important to have reliable estimates of the system parameters, notably the inclination which sets the amount of limb-darkening and Doppler boosting (and visible area from which the flux is emitted), the BH mass, which shifts the peak temperature, sets the Eddington limit (and in extreme cases can influence f_{col} through changing the density) and the distance which sets the luminosity (and therefore Eddington ratio). As the luminosity scales as the inverse square of the distance, the resulting spin values are highly sensitive to this parameter and obtaining an accurate distance measurement (for instance via radio parallax measurements: see Miller-Jones [215]) is critical.

As we will discuss in Sect. 3.3.3, the above models have been widely utilised in estimating the spins for BHBs. However, whilst lauding their successes, it is also important to consider the limitations of such models. There are a number of key assumptions that go into both KERRBB and BHSPEC including the assumption that the discs are steady-state (i.e. time independent) and that the flows are radiatively efficient. The first assumption breaks down should winds be driven from the innermost disc by thermal, MHD processes or radiative pressure (e.g. Ponti et al. [259]) as the mass accretion rate is then a function of time and radius. The second assumption is likely to be invalid at very low mass accretion rates where the flow is low density and hot material is carried into the BH before it can radiate, i.e. advected [238, 239]; such flows are referred to as advection-dominated accretion flows (ADAFs) or radiatively inefficient accretion flows (RIAFs). Such ADAFs/RIAFs also appear at very high mass accretion rates where the scale height of the disc is very large and so photons are effectively trapped as the diffusion timescale is longer than the viscous infall timescale [2]. The effect of advection at high accretion rates is to stabilise the disc (i.e. removes heat from the flow) with a change in the emission profile from $T \propto R^{-3/4}$ (see Eq. 3.27) to $T \propto R^{-1/2}$. Nonrelativistic models (with multicolour discs) with a free emissivity profile are available (e.g. DISKPBB: [120, 163, 164, 216, 337]); however, until recently, these did not include the complex calculations necessary to describe a physically motivated disc atmosphere and relativistic transfer/ray-tracing important in the Kerr metric.

The model SLIMBB ([293, 316] utilises ray-tracing and the radial and vertical profile solutions given in Sadowski et al. [293, 294]. This model once again does not account for mass loss in a wind but accounts for three key components in high mass accretion rate ADAFs, the radial advection of heat and subsequent change in the emissivity, the position of the inner edge (which can move from the marginally stable to marginally bound orbit: [2]) and the location of the effective photosphere. A later version of this code, SLIMBH [317], incorporates the TLUSTY stellar atmosphere code directly and so is closer in nature to BHSPEC.

Unlike the case for high mass accretion rate ADAFs, the emission from low accretion rate ADAF/RIAFs may arise from synchrotron cooling in radiatively inefficient jets [88] or strong outflowing winds [28], Bremsstrahlung emission or Compton scattering in the plasma. Whilst the exact nature of the emission is still debated (although correlations between radio and X-ray luminosity may tend to favour emission from a jet, e.g. Corbel et al. [45]; Gallo et al. [95]), it is clear that the optically thick disc is not present, and therefore, such accretion states are not presently used to diagnose the spin.

There is also a more fundamental assumption that goes into models that derive from the Novikov–Thorne [246] prescription: real accretion discs will have a finite thickness and will not behave as if razor thin. Paczyński [250] and Afshordi and Paczyński [3] argue for a monotonically decreasing deviation with decreasing scale height for small α. This assertion was later confirmed by calculations [298] and so in these limits, the GRAD models should be reliable. However, this does not account for the presence of magnetic fields in the disc which are expected to be generated via a dynamo effect and give rise to magnetic stresses and angular momentum transport [14]. Recent 3D GRMHD simulations [243, 244, 257, 299] estimate that both the luminosity and stress in the inner regions differ substantially (by up to 20 %: [244]) to that expected in the Novikov–Thorne prescription. Both Kulkarni et al. [166] and Zhu et al. [345] explore how this might affect estimates for the spin derived from the use of codes such as KERRBB and BHSPEC. The former obtain the disc flux profile for a series of spin values resulting from the 3D GRMHD thin disc simulations of Penna et al. [257], setting $f_{col} = 1.7$ (as with KERRBB), arguing that the extra sophistication of calculating the position of the effective photosphere is unnecessary in this instance. The spectra themselves are then determined via ray-tracing (see the discussion on KERRBB) without returning radiation but taking into account limb-darkening. Zhu et al. [345] perform a similar set of GRHMD simulations but, by including TLUSTY and radiative transfer, also include the distortion to the spectrum as a result of spectral hardening (as described above). Both sets of authors find that unlike Novikov–Thorne discs, emission can be seen to originate from within the ISCO due to a combination of advection at high accretion rates (as discussed above: [295]) and nonzero inner torque, resulting from a finite disc thickness and leading to increased viscous dissipation at radii close to the ISCO. The combination of these two effects leads to the emission peaking at smaller radii than Novikov–Thorne discs giving both a higher peak temperature and larger emitted flux (and which extends inside the plunging region: [345]). By fitting the simulated spectra with KERRBB [166] and BHSPEC [345], the spin is found to be systematically overestimated as a result of the increased disc brightness. Crucially, the typical error resulting from the use of the Novikov–Thorne profile is far less than the errors associated with the system parameters of mass, inclination and distance (see Gou et al. [113]; Steiner et al. [307]); the spin values derived from GRAD-based models can therefore be treated as representative where the errors on the system parameters dominate.

As an addendum, it is also possible that the disc emission profile differs dramatically from all of those mentioned thus far. Novikov–Thorne (and Shakura–Sunyaev) discs are both thermally [302] and viscously [173] unstable, whilst MHD simulations carried out by Hirose et al. [121] find that small patches of the disc can be thermally stable yet viscously unstable. As a result (and motivated by both spectra and variability arguments), Dextor and Agol [63] proposed a toy model for an inhomogeneous disc (ID) which consists of zones which undergo independent fluctuations driven by radiation pressure instabilities and leads to random walks in the effective temperature at that radius. As Dexter and Quataert [64] point out, this model can explain the spectral and variability properties of the soft states in BHBs. Importantly, should this model be an accurate depiction of the disc and its emission, there are implications for our ability to measure the spin from fits to the continuum. The ID model assumes a magnitude of temperature fluctuations, σ_T (not to be confused with the Thompson cross section); increasing this value leads to a higher characteristic temperature at the inner edge (see Eq. 3.27). As a consequence, deriving the position of the ISCO from the temperature using the Novikov–Thorne disc profile could be misleading, resulting in systematic overestimates for the spin (as with the consideration of a finite thickness disc discussed above). Dexter and Quataert [64] quantify the likely effect, finding that the impact depends on the model used to account for the hard tail accompanying the disc emission [192]. In practice, they determine that the errors can be far larger than statistical uncertainty or systematic uncertainty resulting from not considering emission from within the plunging region [166] except in the most disc-dominated states (with a disc fraction $\gtrsim 0.95$ or $\sigma_T \lesssim 0.15$).

3.3.2 Modelling the Reflection Spectrum

A hard tail of emission ubiquitously accompanies the thermal disc emission in BHBs (and AGN—see, e.g. Jin et al. [138, 139]) and must originate by inverse Compton-isation in a corona of thermal/nonthermal plasma in some, as yet, undetermined geometry (e.g. Liang and Price [172]; Haardt et al. [118]). The exact spectral shape (and properties such as variability) of this component is discussed in detail elsewhere in this compilation but can very broadly (though not always accurately) be described by a power law which breaks at the peak temperature of the thermal electron distribution.

In all models, the corona producing the power law emission has a larger scale height than the disc, and as a consequence, the disc will subtend some solid angle to the upscattered seed photons which will re-illuminate the disc down to a typical scattering surface at $\tau_{eff} = 1$ (e.g. Guilbert and Rees [117]; Lightman and White [174]; George and Fabian [107]). The resulting 'reflection' spectrum is composed of scattered re-emission (i.e. Compton scattering in the surface layers), bound-free edges, bound-bound absorption and emission lines. The three most important components of the spectrum are the Fe K_α edge, the Fe K_α fluorescent line(s) and the Compton downscattered hump. The details of these components are discussed in detail in the review of Fabian et al. [85] with only a brief overview provided here.

As Fe is cosmically abundant and the fluorescent yield (the probability that a fluorescent line is produced following photoelectric absorption) of neutral elements scales as Z^4, emission from Fe is expected to be of great importance. Moderate energy (a few keV) X-ray photons produced via inverse Compton scattering in the corona are energetic enough to remove the inner K (1s) shell electron (leading to a sharp photoelectric edge at 7.1 keV). As long as electrons are available in the L (2s) shell, one will drop to fill the K shell gap and release a photon at either 6.404 keV ($K\alpha_1$) or via spin–orbit interaction, a secondary transition at 6.391 keV ($K\alpha_2$). The probability of this occurrence is only 34%, whilst the most favoured (66% probability) outcome is Auger de-excitation where the production of a photon via the $L \rightarrow K$ shell transition is absorbed by a bound electron which is subsequently expelled. These energies assume that the Fe is neutral; however, illumination of the disc will act to increase the ionisation state of the reflecting material (e.g. Ross and Fabian [283]; Ross et al. [284]), parameterised as $\xi = L/nr^2$, where n is the electron density and r is the distance from the ionising source of luminosity L [322]. An increase in ionisation state increases the electron binding energy and the energy of the edge and Fe K fluorescent doublet accordingly; however, the lines only emerge significantly above 6.4 keV for Fe XVII and above. The fluorescent yield will also (weakly) depend on ionisation state, and when Fe is Li- through to H-like, Auger de-excitation is no longer possible as two L shell electrons are required. Instead, photoelectric recombination can lead to line emission (with a high fluorescent yield) at $\sim$6.8 keV.

Although Fe K is the strongest feature of the reflected emission, other metal transitions (e.g. Ni K_α) also contribute to the overall picture through line emission and absorption edges. Thus, the precise details of the reflected emission clearly depend on the elemental abundances and ionisation state of the illuminated material and have been discussed in detail by Matt et al. [189, 190] who consider illumination of a constant density slab. They find four distinct regimes of increasing ξ, ranging from 'cold' reflection for ξ < 100 ergs/cm/s (where reflection around Fe K resembles that expected from neutral material and the absorption edge is saturated and weak) and terminating at ξ > 5000 ergs/cm/s where there is no absorption edge or line. In between these regimes, the strength of the line is dependent on the number of electrons available in the L shell as described above, with a weak line and moderate edge produced by ionised species of Fe up to Fe XXIII (where the Auger effect may still take place) and a stronger 'hot' line by species above FeXXIII (where the Auger effect no longer takes place and line emission is a result of recombination). In the latter case, the edge appears stronger due to increased flux below the edge as a result of diminishing opacity. One of the most successful models which accounts for illumination onto a semi-infinite slab of optically thick material in the atmosphere of an accretion disc is REFLION and its later incarnation, REFLIONX [284, 285]. These models, based on the work of Ross et al. [287] and Ross [286], fully incorporate the radiative transfer of continuum X-rays (using the Fokker–Planck diffusion equation), line emission and Comptonisation (using the modified Kompaneets operator) and thus were an important step forward from earlier models which, whilst not including line emission or intrinsic emission from inside the gas, were the first to incorporate Green's functions to describe the scattering of photons by electrons in cold gas (e.g. PEXRAV: [179]).

The important effects in creating the reflection spectrum discussed so far make no mention of the effect of spin, but this has a significant impact on the emergent spectrum for a number of reasons. Principally, the Fe K_α emission line (s) originates from illuminated disc radii which rotate in circular Keplerian orbits (although deviations are expected to scale with $(H/R)^2$). The observed reflection spectrum is a composite of emission from the receding Doppler red-shifted and approaching blue-shifted sides which leads to separation in line energies and a classic 'double-horned' profile; as these radii approach the BH, the radial velocity increases which leads to a greater separation. As the orbiting material near the BH is mildly relativistic, beaming of the emission (resulting from relativistic aberration and time dilation) leads to a flux change of a factor D^3 where D is the Doppler factor:

$$D = \{\Gamma[1 - \beta\cos(i)]\}^{-1} \tag{3.34}$$

Γ is the Lorentz factor, $\beta = v/c$, v is the approaching or receding velocity and i is the inclination of the observer from the rotation axis. As a result of relativistic beaming, flux from the approaching, blue-shifted side is Doppler boosted, whilst the receding side is Doppler deboosted (see, e.g. Fabian et al. [82, 85]; Stella [312]). Accompanying these Newtonian and special relativistic effects are the special relativistic effect of transverse Doppler shift and the general relativistic effect of gravitational redshift, both of which act to reduce the observed energy of the line at each radius. It can be seen that the combination of these effects leads to a heavily skewed and broadened line, where the blue wing depends heavily on the inclination (Eq. 3.34) and the shape of the red wing is dominated by the position of the inner edge and can therefore be used as a proxy for the BH spin (e.g. Laor [167]). In Fig. 3.3, we demonstrate these effects on a fluorescence line originating from two annuli in the

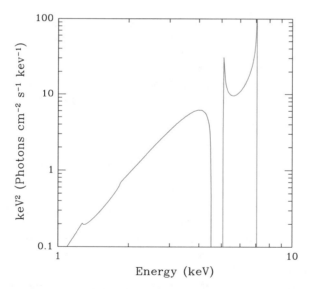

Fig. 3.3 Distortion of an emission line (using the model RELLINE: [54]) for two annuli in the accretion disc seen at 45 degrees about a BH with $a_* = 0.998$. The *red line* is from an annulus 20–22 R_g from the BH and is broadened into the characteristic 'double-horned' profile by Doppler shifting. The *blue* and *red* 'wings' of this line are then increased and decreased in flux, respectively, through the effect of Doppler boosting (relativistic beaming: see Eq. 3.34). The *blue line* comes from an annulus at the ISCO (1.3-2 R_g from the BH) where the *red* wing gives a measure of the position of the ISCO; the entire line is shifted to lower energies due to the combination of transverse Doppler shift and gravitational redshift. In reality, we see a blend of lines from all radii

disc. Importantly, unlike the case with continuum fitting, the method of fitting the relativistically broadened Fe line is *independent* of BH mass. In addition, the technique is insensitive to the distance to the source and the inclination is a ubiquitous parameter of the spectral models (determined from the shape of the blue line and Fe K_α absorption edge) and so can be estimated concurrently with the spin. As a consequence, the major source of error when determining the spin is the systematic uncertainty within the model itself. Of importance when determining the emergent reflection spectrum is the modelling of the primary illuminating continuum which may be more complex than a simple power law but can be probed through the use of an expanded bandpass (e.g. the *NuSTAR* observations of Cyg X-1: [255]) as well as through the application of advanced spectral timing techniques (see the review of Uttley et al. [327]). Additionally, the radial and vertical dependent density structure of the optically thick disc and the geometry of the corona/disc system (which sets the emissivity, e.g. Ghisellini et al. [108]) are important considerations for accurate modelling of the reflected emission. The following subsections discuss these issues in turn.

3.3.2.1 Emissivity and Geometry

An important ingredient of the reflection spectrum is the 'emissivity' (ϵ) of the flux from the disc which is proportional to the radial dependence of the illuminating radiation *onto* the disc (where the irradiation goes as $I(R) \propto R^{-\epsilon}$). For an irradiating point source in flat space-time, sitting above the disc, the emissivity goes as the product of the inverse square law and the cosine of the normal to the plane of the disc (see the discussion in Wilkins and Fabian [339]) or, for a source, h above the disc, ϵ goes as $(R^2 + h^2)^{-1} * \cos\theta$ (where $\cos\theta$ is just $h/\sqrt{(R^2 + h^2)}$). Thus, at small radii ($R \ll h$), this would tend to a flat profile but at large radii (around $R = h$) will instead tend to R^{-3} which is that of the disc emission in Eq. 3.27. This discounts the effect of relativity however, and the effect of light-bending close to the BH can have a significant impact by focusing more of the radiation onto smaller radii.

The impact of light-bending naturally depends on the location of the corona and the geometry of the disc–corona system. One of the most popular geometries is that of the 'lamp-post' (see Fig. 3.4) where the corona sits on the BH spin axis some height above the BH; in a physical sense, this would then be associated with the base of a jet (see Markoff and Nowak [184]; Dauser et al. [53]; Wilkins and Fabian [339]; Wilkins and Gallo [340]). The result of light-bending in this geometry can lead to highly anisotropic illuminations and is predicted to produce a radial profile that is a twice-broken power law, with very steep emissivities in the most inner regions then flattening before tending to constant emissivity at large radii [188, 217]. Naturally, as the profile is a function of the light-bending, it is a function of the source height above the disc. For a decrease in height, the emissivity profile is steepened due to an increased amount of light-bending, i.e. a larger value of ϵ [218, 273, 339], which drops the observed flux from the corona and increases the amount of reflection, the ratio of which is the 'reflected fraction'. In addition, the position of

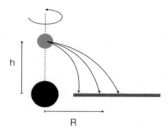

Fig. 3.4 Lamp-post geometry showing the X-ray source (*blue*) at some height h above the BH (on the spin axis) as expected for the base of a jet. The X-ray photon paths are bent by the strong gravity and illuminate the disc, leading to a steeper emissivity profile at small radii than a simple flat profile (which breaks to R^{-3} at large disc radii) otherwise expected from illumination in flat space-time

the break to a flatter emissivity changes as does the extent of the region over which this is predicted to hold [339]. Although reflection models without ray-tracing (or incorporating a relativistic transfer integration kernel) cannot account for this effect *directly*, convolution models which take into account the effect of photon orbits in the Kerr metric have been developed and can be used to alter the emissivity profile accordingly. The most commonly used models (besides RELCONV which we will mention shortly in the context of next-generation models) are KDBLUR/KDBLUR2 which are based on the relativistic transfer function of Laor [167] and KERRCONV which uses an analytical prescription for relativistic beaming and a pared down transfer function for ease of computing [32].

The geometry of the illuminating source remains a fundamental issue in accretion physics in general but has especially important relevance for reflection models (although as we will discuss, there are time domain methods now in use to help constrain this) as this sets the possible emissivity profiles. It may well be lamp-post-like in the picture where the optically thin electron population is associated with the base of a jet (and, if the component is analogous to that in BHBs, probably composed of a thermal electron population). In the case where the scattering is not in a jet base, other geometries must be considered. Wilkins and Fabian [339] explored a series of potential geometries for the corona, notably testing not only for position of the illuminating source but also for the source's extent which had not been previously considered from a theoretical perspective. They calculated emissivity profiles from ray-tracing using GPUs (graphics processing units) and included a full treatment of the relativistic effects including the reduction in effective disc area (which they find goes as the classical disc area divided by the redshift: see their Fig. 3.2) and blue-shifting of photons onto the disc. The geometries considered include an axial source (lamp-post geometry) both stationary and moving away from the BH (i.e. a jet), an orbiting source (which can be seen as a single element in a ring) and an extended disc-like source. In the case of the ring source, the act of moving it further from its position above the BH leads to a somewhat shallower emissivity profile in the central regions with a flatter profile after the first break and extending to larger radii, dependent on the radial position of the ring. As there are several arguments for an extended corona from both variability studies (e.g. Churazov et al. [44]) and more directly from gravitational microlensing (which indicates an extension of up to a few R_g: [52]), an isotropic point source is unlikely to be representative, and to this end, the study of an extended source is extremely valuable. Wilkins and Fabian [339] describe such a scenario as the sum of points spread over a vertical (giving a source of finite thickness) and radial extent (i.e. a disc above the disc). The source is assumed to have a constant luminosity across its extent and is optically thin to the radiation it has produced (and so does not interact with itself); as expected, the resulting emissivity profile is a combination of the breaks and indices from the consideration of point sources (although may well be further complicated if magnetic reconnection occurs at distinct radii in the disc, e.g. Sochora et al. [305]). Wilkins and Fabian [339] explore the effect of changing the spin on the emissivity, finding that the increased disc area resulting from a higher spin leads to a steeper emissivity for an axial or ring geometry (with [339] showing that the emissivity is only greatly steepened in these arrangements for $a_* \geq 0.8$).

Finally, in addition to the issues associated with the as-yet-unknown coronal geometry, a further simplifying assumption commonly made is that the coronal plasma itself is uniform, being of a single temperature and density. This is unlikely to be the case (see, e.g. Parker et al. [255]) and in future can be tested using eclipses to determine the radial dependence of the coronal properties (see Sect. 3.5).

3.3.2.2   Disc Vertical Density Profile

As with estimating the BH spin via the disc emission (see previous section), the reflection spectrum also depends on the vertical structure of the disc; in the former, the important consideration was the position of the effective optical depth (which, as we discussed, can change depending on the relative dominance of opacities). In the case of reflection spectra, the important consideration is the ionisation state of the material in which the majority of the reflection takes place (i.e. one effective optical depth). Shakura and Sunyaev [301] point out that in radiation pressure-dominated discs of large optical depth and heated by viscous dissipation, the density in the vertical direction is roughly constant; this has led to models for both AGN [190, 283, 350] and BHBs [284, 288] incorporating this simplifying assumption of a constant vertical density structure in the disc (or a Gaussian distribution: [284]). This assumption has been questioned by Nayakshin et al. [237] who point out that the illuminating X-ray heating of the outer disc layers (where reflection occurs) could be orders of magnitude greater than the viscous heating in the disc at such height above the mid-plane. The disc material is expected to be thermally unstable when the supporting gas pressure in the illuminated atmosphere falls below a fraction of the downward radiation pressure provided by the illuminating continuum [157, 162, 265, 289, 290] which leads to a large jump in temperature. Nayakshin et al. [237] argue that a self-consistent determination of the density gradient is therefore necessary and determine the density profile from the condition of hydrostatic equilibrium, simultaneously solving the equations of ionisation, energy balance and radiative transfer, finding constant density (or Gaussian distributions) to be broadly unphysical. Their calculations instead suggest that the structure of the illuminated layer of the disc is a two-phase structure formed of an ionised skin where, in the case of 'hard' illuminating flux, Fe is completely ionised, and below this layer, the material is cold, i.e. weakly ionised. The strength of the reflection features is then a function of the optical depth of the top layer (which does not imprint features due to being totally ionised), with smaller optical depths of the skin leading to less of an impact on the emergent reflection spectrum from the cold layers beneath (as such layers are reached more readily). This in turn depends on the strength of the illumination and is therefore intrinsically connected to the emissivity profile discussed above. As pointed out by Fabian and Ross [84], simulations of discs supported by magnetic pressure (e.g. Blaes et al. [27]) add a complicating factor, as this can potentially reduce the density and increase the impact of photoionisation. Indeed, should magnetic pressure dominate the hydrostatic support, then the discontinuous vertical profile resulting from the thermal instability may not exist, although large-scale inhomogeneities could be present and these could have an impact on the emergent reflection spectrum [16, 27]. As the impact of magnetic fields on the disc structure remains an open question, so too does the nature of hydrostatic balance.

3.3.2.3 Effect of the Plunging Region

A standard assumption in modelling the emission from the BH accretion flow is that truncation at the ISCO is final with no radiation arriving to the observer from the ballistically infalling material in the plunging region. Zhu et al. [345] studied this low-density region through 3D MHD simulations, discovering that only a small amount of thermal emission was produced and was unlikely to distort spin values obtained via the continuum fitting method (Sect. 3.3.1).

In order to determine the effect of including the plunging region on the *reflected* spectrum (as first discussed in Reynolds and Begelman [273]), Reynolds and Fabian [272] performed high-resolution 3D MHD simulations (thereby incorporating MRI-driven turbulence: [14, 15]) of the geometrically thin accretion disc close to the ISCO. Their simulation demonstrates that some Fe K_α emission can originate from the plunging region which introduces an additional systematic error for models which terminate at the ISCO. As a notable caveat to this effect, at high mass accretion rates the density in this region drops and the material may become ionised to the point where there are no longer any notable reflection features (with the exception of the Compton hump). The authors estimate the uncertainty introduced by the inclusion of reflection from within the plunging region (see their Fig. 3.5), finding it to be an overestimate of the true spin (as for the inclusion of emission from this region in continuum fitting, e.g. Zhu et al. [345]) with a larger error for slowly spinning BHs, and to be relatively insensitive to uncertainty in the inclination (as long as the inclination can be constrained from modelling the Fe K_α profile). The uncertainty grows when the scale height is large just outside of the ISCO; whilst further work is needed to establish the full impact of this, Reynolds [271] notes that inclusion of

Fig. 3.5 Schematic of a wind showing a gap (assumed to have been formed via Rayleigh–Taylor or similar instabilities), rotating and providing a changing view through to the inner disc regions (the emission from which is determined from ray-tracing: [62]). Analogous to the situation of eclipses by Compton-thick clouds [281], the emission seen by the observer as a function of time is dependent on the inclination and spin (as well as disc structure). By incorporating the additional lever arm of variability, Doppler tomography provides a means to obtain tighter constraints on the spin than is possible through the use of the time-averaged spectrum alone [211]

such uncertainties can relax the constraints for otherwise maximal spinning BHs to >0.9.

3.3.2.4 Next-Generation Models

The most recent codes (developed principally by Thomas Dauser, Javier García and colleagues) provide important updates to existing models. In the case of nonrelativistic reflection (i.e. no light-bending effects), XILLVER [99–101] includes XSTAR [142] to solve for the ionisation state of the disc atmosphere and makes use of the most updated, accurate and complete atomic database for atomic transitions. A series of codes have also been developed which include a relativistic transfer function from ray-tracing which allows the emissivity of the irradiation to be determined (although the disc is still assumed to be 'thin' as in the Novikov–Thorne prescription). These models include RELLINE [54], which determines lines emission for spins ranging from maximal retrograde through to maximal prograde and incorporates limb-darkening, limb brightening and is similar in output to previous models including KERRDISK and KYRLINE [71] but is evaluated over a finer energy grid than LAOR. This model has been adapted to act as a convolution kernel (RELCONV) for the entire spectrum which, when combined with XILLVER, allows the entire reflection spectrum to be calculated with a prescribed emissivity law (RELXILL: [53]). Using RELXILL, Dauser et al. [55] showed that by considering the additional spin dependence of the reflected fraction (which most models do not account for), it becomes possible to place increasingly stringent constraints on the spin by discounting unphysical solutions (see also Parker et al. [254]), although consideration of distant, neutral reflection may complicate matters somewhat. In addition to improvements in determining the emissivity, RELXILL also allows the thermal cut-off (at energies often well beyond the detector bandpass) to be determined [103] whilst calculating the angular dependence of the reflection spectrum (approximated in the past by a convolution of the angle-averaged reflection spectrum with a relativistic kernel) in a fully self-consistent manner. The latter has only a minor impact on measurements of the spin and inner disc inclination (although the constraints on both parameters improve) but can have a major impact on the estimated Fe abundance [102]. Although an undoubted step forward, this model still relies on several caveats that are important to consider; the radial and vertical density profile is assumed constant, as is the ionisation state of the reflecting material. Further iterations of the RELXILL[1] model have begun to deal with these assumptions, by considering a power law-like radial density structure, multizone ionisation structure or with the ionisation gradient calculated self-consistently from irradiation.

3.3.2.5   The Geometry Through Reverberation

As the geometry of the disc–corona system is potentially important for reliably estimating the spin, we will very briefly discuss the role of reverberation as the most promising method for placing tight (spectrally independent) constraints.

It was realised early on (Fabian et al. [82], Stella [312]) that the physical separation of the corona and the accretion disc will, from the perspective of an observer at infinity, lead to a light-travel time delay between the emission arriving directly from the corona and that reflected from the disc (see Matt and Perola [191]; Campana and Stella [39]; Fabian et al. [85]); a simple schematic showing this for the lamp-post geometry is given in Fig. 3.4. This 'reverberation' off the disc can be studied in the time domain from the cross-correlation function (e.g. Gandhi et al. [98]) between a band containing the intrinsic hard emission (from the corona) and a band dominated by reflection. For ease of evaluating the lag across multiple frequencies, reverberation is more commonly studied in the Fourier domain via 'phase lags'. As the phase lag is a function of the geometry, it can provide important insights into the nature of the accretion flow and has implications for how we measure the spin. In practice, obtaining the phase lag requires evaluating the components of the cross-spectrum [247, 331] which in turn are found from the frequency-dependent complex ordinates of the Fourier-transformed light curve. Here, we present only the most basic of descriptions; for a detailed review of Fourier analysis, we point the reader to van der Klis [329], and for a detailed review of reverberation and phase lags, we recommend that of Uttley et al. [327].

The light curve in each energy band (e.g. those dominated by either the primary or reflected emission) is given by $x(t)$ and $y(t)$ with their Fourier transforms $X(\nu)$ and $Y(\nu)$. An alternative way of displaying these is $X(\nu) = A(\nu)e^{i\Psi}$ and $Y(\nu) = B(\nu)e^{i\Psi+\phi}$ where ϕ is the phase lag between them. By defining the cross-spectrum as

---

[1] www.sternwarte.uni-erlangen.de/~dauser/research/relxill/.

$C(v) = X^*(v)Y(v)$ (Nowak and Vaughan [245]), the phase can then be determined from:

$$\phi(v) = \arg[C(v)] = \tan^{-1} \frac{\Re(v)}{\Im(v)} \qquad (3.35)$$

where $\Re$ and $\Im$ are the real and imaginary Fourier coefficients of the cross-spectrum at each frequency. The details of how to estimate the phase in practice (i.e. taking averages over segments and normalising) are discussed in Uttley et al. [327]. The important question is how do we go from the phase as measured from the light curve to the geometry? Each light curve can be related via an impulse response to an input light curve, called a 'driving signal', $s(t)$, such that:

$$x(t) = \int_{-\infty}^{\infty} h(t - \tau)s(\tau)\mathrm{d}\tau \qquad (3.36)$$

where τ is the time lag ($\phi/2\pi v$). In Fourier space, this is equivalent to saying $X(v) = H(v)S(v)$ (and similarly, $Y(v) = G(v)S(v)$), where $H(v)$ (or equally $G(v)$) is the Fourier transform of $h(t - \tau)$ and is called the transfer function. As described in Uttley et al. [327], the impulse response (and therefore the geometry) is related to the phase lag in Eq. 3.35 via the cross-spectrum:

$$C(v) = H^*(v)S^*(v)G(v)S(v) = |S(v)|^2 H^*(v)G(v) \qquad (3.37)$$

Thus, the cross-spectrum of two light curves contains the cross-spectrum of the transfer functions ($H^*(v)G(v)$) and the power spectrum of the driving signal. As the power spectrum of the driving signal ($|S(v)|^2$) is by definition real-valued, it has no effect on the phase, and so, given an input relativistic transfer function from the geometries described in Sect. 3.3.2.1, it therefore becomes possible to compare the expected phase lag to observations and thereby begin to constrain potential geometries (e.g. Cackett et al. [37]). In so doing, such analyses also hold the promise of better understanding contributory factors in measuring the spin.

As reverberation in BHBs occurs on very fast timescales, the number of photon counts per light crossing time is small; as such, observations of reverberation due to reflection have to date focused mainly on AGN (where conversely, the number of photon counts per light crossing time is substantial). The first tentative hints of a signal were noticed by McHardy et al. [196] with the first significant ($\geq 5\text{-}\sigma$) detection of a reverberation lag due to reflection made by Fabian et al. [83] from *XMM–Newton* observations of the AGN, 1H0707-495 (see also Zoghbi et al. [346]). Due to the remarkably strong Fe L emission in this source (which is noted for having supersolar abundance of Fe), the authors were able to take a hard band without strong contributions from reflection and detect the signature of reverberation as a 'soft lag', i.e. the soft emission lagging the hard. Notably, the strong Fe K$_\alpha$ emission seen in the energy lag spectra (e.g. Zoghbi et al. [347, 348]; Kara et al. [143, 145, 146]) and its continuation to higher energies mapping out the Compton hump (through the use of *NuSTAR* data: [144, 349]) on the frequencies of the reverberation lag all indicate that its origin lies in relativistic reflection.

Following the initial discovery of reverberation, a soft lag has now been detected (or hinted at) in several AGN (e.g. Emmanoulopoulos et al. [77]) with the detection of a significant trend of frequency/amplitude with mass [60]. This would require that the coronal geometry be the same throughout, although in at least one AGN (IRAS 13224-3809: [147]) the soft lag is observed to change in frequency and amplitude; this does not necessarily invalidate a correlation, but understanding the stability and evolution of the corona–disc geometry is clearly of great importance.

Work is now turning towards reconstructing the spin-dependent geometry (see Sect. 3.3.2.1 and Wilkins et al. [340] for a recent study of this) through the use of impulse response functions (e.g. [274]) and *direct* modelling of the phase lags. Young and Reynolds [342] and more recently Cackett et al. [37] and Emmanoulopoulos et al. [78] simulate the effect of reverberation assuming a lamp-post geometry and obtain the associated impulse response functions for a variety of physical parameters (either via precalculated transfer functions in the case of the former or via direct ray-tracing in the case of the latter). The major effect on the frequency-dependent lag results from changing the mass or vertical displacement of the axial illuminator and is relatively insensitive to the inclination and spin. The energy profile of the lag, however, is far more sensitive, with the profile of Fe K_α—found to be strong at frequencies corresponding to the soft lag—changing with spin in a similar manner to its time-averaged counterpart. Cackett et al. [37] compare predicted frequency-lag and energy-lag spectra to the real data of NGC 4151 and constrain the physical parameters including the spin (found to be maximal), inclination and source height. In a similar approach, Emmanoulopoulos et al. [78] perform the first sample analysis, modelling the frequency-lag spectra in 12 AGN using general relativistic impulse response functions, finding a consistent source height across the sample and a possible bimodality of spin values (which might indicate SMBH growth via mergers, e.g. Volonteri et al. [332]). In order to reliably estimate the spin, these techniques typically rely on high-quality data with high energy resolution; whilst not widely available at present, the arrival of next-generation missions such as ESA's *Athena* will allow such methods to be used to their full potential. In addition, *Athena's* location at the L2 point will allow uninterrupted observations which, in turn, will allow lower frequencies to be explored, important for larger mass AGN where the reverberation signal is expected to be found (e.g. De Marco et al. [60]).

3.3.3 Results: BHBs

Here, we will discuss the results of campaigns to estimate the spin of stellar mass BHs from applying the methods described in Sects. 3.3.1 and 3.3.2. Due to their success, the number of sources for which spin measurements have been obtained is constantly growing and we apologise for any results which are therefore absent. Whilst we briefly touch upon the relevance of measuring the BH spin for the purposes of probing the launching of astrophysical jets, obtaining an understanding of the spin distribution is important in its own right as this provides a view of the natal spin which is set during the supernova process (as insufficient mass can been accreted onto the BH to change the spin by a considerable amount: see the discussion of Reynolds [271] and Sect. 3.2).

3.3.3.1 Continuum Fitting

As discussed elsewhere in this compilation, the outburst of BHBs (with low-mass companion stars) follows a predictable path through X-ray spectra, variability and multiwavelength properties [89, 90]. The luminosity of a source is usually parameterised as a ratio to the Eddington limit for spherical accretion, given by:

$$L_{Edd} = \frac{4\pi\, GMm_p}{\sigma_T} \tag{3.38}$$

where σ_T is the Thompson cross section and m_p is the proton mass. This is frequently referred to by its numerical approximation (and under the assumption that the material is entirely ionised Hydrogen), $L_{Edd} = 1.26 \times 10^{38}$ $M_{BH}/M_\odot$.

On the rise to outburst, at low-to-moderate mass accretion rates (typically <70% of the Eddington limit: [73]), the spectrum is dominated by a hard tail of emission resulting from inverse Compton scattering by an optically thin, thermal population of electrons (thus, the spectrum is referred to as being in a hard state). This emission is accompanied by synchrotron emission extending across several decades in frequency and is associated with a low bulk Lorentz factor (Γ) jet. At the highest mass accretion rates, the spectrum begins to soften and becomes increasingly dominated by emission from the accretion disc, passing through the intermediate states and then into the soft state (we note that there are accompanying changes in the variability properties [24], but these are not of relevance for the discussion here).

The soft state (also called the thermal dominant state) is characterised by a strong disc component and a nonthermal (rather than thermal) tail of emission to high energies which is only of the order of a few per cent or less of the total flux in the X-ray bandpass (e.g. McClintock and Remillard [192]; Remillard and McClintock [270]). As the vast majority of the emission originates from the disc (with very little energy being liberated in a corona), the spin can in principle be determined from the application of a suitable model for the disc emission (see Sect. 3.3.1). However, a condition of applying this technique relies on the inner edge being located at the ISCO (as expected from GRMHD simulations: [257, 299]). As the luminosity of the disc component for a fixed emitting area (i.e. fixed inner edge) is expected to follow a T^4 dependence (see Eq. 3.27), this provides a means by which to test the consistency of the position of the ISCO. Determining the position and stability of the ISCO via this approach has been attempted by a number of authors (e.g. Ebisawa et al. [74]; Muno et al. [233]; Kubota and Done [165]; Kubota and Makishima [164]; Steiner et al. [307]). In particular, Gierliński and Done [110] use the expected relation for the integrated disc luminosity as a function of the maximum observed colour temperature [111] considering a pseudo-Newtonian potential (Paczyńsky and Wiita [251]) and corrections to the observed flux due to inclination and GR effects [48, 344] to obtain the predicted form of the relation between Eddington ratio (L/L$_{Edd}$) and temperature. Through the use of *RXTE* data (which covers a nominal 3–20 keV energy range) for a number of well-known BHBs, Gierliński and Done [110] showed that the disc emission in the soft state is consistent with a fixed inner edge for a range of Eddington ratios, with deviations at the very highest and lowest values. As the predicted relation depends on GR corrections and therefore the spin, Gierliński and Done [109] were able to use these plots to indicate that the spin in the case of XTE J1550-564 is non-maximal.

The first attempts to *directly* constrain the spin from modelling the disc emission (and indeed pioneering the field of continuum fitting) were carried out by Zhang et al. [344]. The authors utilised values for the peak temperature and flux for a number of BHBs available from the literature (e.g. Dotani et al. [70]; Belloni et al. [22]) to obtain the position of the inner edge after accounting for relativistic effects. Following the development of new models (namely KERRBB and BHSPEC), there has been a steady and substantial increase in spin measurements obtained via this method, with 12 in total covering persistent and transient sources both within our Galaxy and in nearby galaxies. McClintock et al. [194] provide a review of the continuum fitting method and its application up to 2013, and we highly recommend this as an excellent overview.

As mentioned in Sect. 3.3.1.1, of critical importance in all attempts to constrain the spin via the continuum fitting method, is the accuracy with which the system parameters—the inclination, distance and BH mass—can be determined as the uncertainty on these dominates over typical model systematics of ∼5 %. As noted by Orosz et al. [248], having an accurate distance to the system is key to reducing uncertainty on the mass estimate. For those systems in nearby galaxies, cosmic distance ladders can be used and result in ∼a few per cent uncertainty, substantially better than distance measurements for Galactic sources [141], although these are being substantially improved through radio parallax measurements (see Miller-Jones [215]). Once the distance is known, the mass can be estimated from modelling the orbital dynamics of the system using eclipsing light curve (ELC) models (e.g. Orosz and Hauschildt [249]; Orosz et al. [248]) and requires the radial velocity of the companion star derived from line spectroscopy. Such modelling also determines the inclination of the system; however, the question remains as to whether the system inclination is representative of the inclination of the inner disc which is key for determining the spin via the continuum fitting method. As discussed in Sect. 3.2.1.1, misalignment of the BH spin axis with that of the binary orbit can lead to relativistic precession and the Bardeen–Peterson effect which aligns the inner regions with the BH spin axis, whilst the outer regions align with the binary plane. There may already be evidence in support of this scenario; the inclination of the jet, which is expected to be the same as the BH spin axis, appears misaligned with the binary orbit in the BHBs, GRO J1655-40 ([114, 122] and also [219] for possible issues associated with accurately determining such physical properties) and SAX J 1819–2525 (also known as

V4641 Sgr: [177, 183]). As an interesting aside, it is possible that the ubiquitous low-frequency quasi-periodic oscillations (LFQPOs) seen in the variability power spectra of BHBs when in the hard through to the intermediate states (see the review of [23]) are associated with precession of a low-density flow in the inner regions due to misalignment [132, 133]. Incorporating the effect of reflection of the radiation from the precessing regions leads to unambiguous, observational tests for misalignment which, if confirmed, would have implications for our ability to measure the spin reliably (see Ingram and Done [130]). We will return to the observational effect of precession in Sect. 3.4.

In obtaining the spin via the continuum method, a further selection criterion is often applied to the X-ray spectra. As discussed in McClintock et al. [193], above $\sim 30\%$ of the Eddington limit, the spectra may deviate from those expected from a simple thin disc possibly due to the creation of an inner, optically thick, radiation-pressure-dominated corona/slim disc due to the high mass accretion rates [2, 205, 206, 326] or a region supported by magnetic pressure [317]. It is assumed that either disc truncation or simply a cooling of the disc photons by the corona leads to a lower disc temperature than should be expected from the innermost edge of the disc and a deviation away from the expected $L \propto T^4$ relation [109]. Such deviations in disc structure can in principle lead to the spin being underestimated and has been proposed to explain the difference in spin results for the extreme BHB, GRS 1915+105 with both moderate [209] and maximal [192] values claimed. However, the actual luminosity at which this distortion appears in GRS 1915+105 is model dependent and requires careful treatment [209]; as a result, the spin of this unusual source remains somewhat contentious. In general, the point at which radiation pressure starts to affect the structure of the flow must be related to the spin (which sets the radiative efficiency—see Eq. 3.30) and so the demarcation at 30% of the Eddington limit translates into differing mass accretion rates for different sources. Should the structure of the disc be affected instead by something coupled to the mass accretion rate, the change in the flow could conceivably occur at a different Eddington ratio; thus, to ensure rigour, selection criteria should ideally be determined on an object-by-object basis (e.g. Steiner et al. [307]).

An important consideration for spectral studies of BHBs (in general but in particular in using the spectrum to estimate the spin) is that whilst their proximity leads to high fluxes, these can lead to severe issues for CCD spectrometers (see [158] for a discussion of these effects on *XMM-Newton* data) including photon pile-up (where the arrival of multiple photons is read as a single event), charge transfer inefficiency (CTI, where a loss of charge occurs during CCD read-out) and X-ray loading (where very bright sources contaminate the 'offset map'—analogous to a 'dark frame' in optical instruments). Such effects can readily distort the spectrum, and therefore, care must be taken to ensure that their impact on estimating the spin is understood.

Finally, it is important to note that mass-loaded winds are launched from the accretion disc (perhaps as a result of radiation pressure, e.g. Proga and Kallman [262]; thermal reprocessing in the outer disc, e.g. Begelman et al. [21], or MHD driving, e.g. Neilsen and Homan [241]) and become stronger as the source becomes dominated by thermal disc emission [259]. Although these winds are highly ionised (as the source is X-ray bright) and are unlikely to present an obstacle for studying the continuum (although see the following section for their impact on AGN spectra), the models applied in order to determine the spin (see Sect. 3.3.1.1) are at present only steady-state and do not take this mass loss into account—the effect of mass loss on the measurement of the spin at this time is therefore unknown.

In Table 3.1, we present the available BH spin values for BHBs (both Galactic and extra-galactic) along with the system parameters and the model used (useful in the light of the assumptions that underpin each model as discussed above). The vast majority of these values are reported in the recent review by Miller and Miller [214], and for the sake of brevity, we direct the interested reader there for the individual references for the spin values and system parameters, although we note, where possible, updated values and any additional sources.

Notably, all bar two of the BHBs in Table 3.1 is Galactic (or located in the large Magellanic cloud), and as such, the issue of precision in the distance measurement is important. In the case of the source in M31 [203, 205], the distance is known to within a few per cent, and as this is relatively large, the source flux (which is high in such soft states) does not pose an issue for CCD detectors. Thus, such bright (yet relatively nearby) extra-galactic sources offer a means to expand our sample of BHBs with spin measurements which, as we discuss in Sect. 3.3.5, may be extremely important for understanding how astrophysical jets are launched.

3.3.3.2 Reflection Fitting

As mentioned in the previous section, due to the proximity of Galactic BHBs, their X-ray flux is typically very high; unsurprisingly, these provided the very first discovery of a broad Fe K_α line through an *EXOSAT* observation of Cygnus X-1 [20, 82]. As the disc emission does not need to be isolated, even higher Eddington fraction observations than those typically selected for the continuum fitting method may be utilised (although should advection occur at such higher rates, then particle orbits may be affected, e.g. Narayan and Yi [238]). Due to the source brightness, CCD detector issues may once again become important, but where these distorting effects can be reliably ignored or corrected (see Miller et al. [212] for a discussion of pile-up on Fe K_α line measurements), CCD spectroscopy of BHBs offers the opportunity to study the reflection spectrum in remarkable detail and thereby well constrain the spin.

As discussed in Sect. 3.3.2, modelling of the reflection spectrum is less sensitive to uncertainties in system parameters such as the distance and BH mass (whilst the inclination is a free parameter in the models), and as such, the method is in principle more robust than modelling the disc emission. However, an important consideration when modelling the reflection spectrum in BHBs is the modelling of the continuum which, in the bandpass of most detectors, contains a significant contribution from the disc due to the BH mass (see Eq. 3.27). In addition, the nature of the Comptonised emission remains a point of debate, with suggestions from broadband spectroscopy extending to high energies, that it may be a combination of more than one component

Table 3.1 Table of BHB spins

Source	Continuum fitting					Reflection fitting
	Mass ($M_\odot$)	Inclination (degrees)	Distance (kpc)	a_*	Model	a_*
Cygnus X-1	14.8 ±1.0	27.1 ± 0.8	$1.86^{+0.12}_{-0.11}$	≥ 0.95	K2	>0.97[a]
XTE J1550-564	9.10 ± 0.61	74.7 ± 3.8	$4.38^{+0.58}_{-0.31}$	$0.34^{+0.37}_{-0.34}$	K2	0.55 ± 0.22
XTE J1650-500						0.79 ± 0.01
XTE J1652-453						0.45 ± 0.02
XTE J1752-223						0.52 ± 0.11
XTE J1908+094						0.75 ± 0.09
A 0620-00	6.61 ± 0.25	51.0 ± 0.9	1.06 ± 0.12	0.12 ± 0.19	K2	
4U 1543-475	9.4 ± 1.0	20.7 ± 1.5	7.5 ± 1.0	0.8 ± 0.1	K	0.3 ± 0.1
4U 1630-472						$0.985^{+0.005}_{-0.014}$
MAXI J1836-194						0.88 ± 0.03
GRO J1655-40	6.30 ± 0.27	70.2 ± 1.2	3.2 ± 0.2	0.7 ± 0.1	K	0.98 ± 0.01
GS 1124-683	7.24 ± 0.70	54.0 ± 1.5	5.89 ± 0.26	$-0.24^{+0.05}_{-0.64}$	K	
GX 339-4						>0.97[b]
GRS 1915+105	14.0 ± 4.4	66 ± 2	11.0	≥0.95	K2	0.98 ± 0.01
				~0.7[c]	B	
GRS 1739-278						0.8 ± 0.2 [d]
SAX J1711.6-3608						$0.6^{+0.2}_{-0.4}$
Swift J1753.5-0127						$0.76^{+011}_{-0.15}$
Swift J1910.2-0546						≤ -0.32
LMC X-1	10.91 ± 1.54	36.4 ± 2.0	48.1 ± 2.2	$0.92^{+0.05}_{-0.07}$	K2	$0.97^{+0.02}_{-0.13}$
LMC X-3	6.95 ± 0.33	69.6 ± 0.6	48.1 ± 2.2	$0.25^{+0.20e}_{-0.29}$	K2	
M31 ULX-2	~10	<60	772 ± 44	<-0.17[f]	B	
M33 X-7	15.65 ± 1.45	74.6 ± 1.0	840 ± 20	0.84±0.05	K	

Notes: Spin values from continuum fitting (and system parameters used for the model fits) and from reflection fitting (see [214] for individual references) with errors typically quoted at 1–2σ (with the exception of the upper limit for the spin of M31 ULX-2 which is at 3σ). The differing levels of quoted precision for the values is a result of the individual studies. The model used in the continuum fitting is either KERRBB: K, KERRBB2: K2 or BHSPEC: B. Updated or additional values are indicated by: [a]Parker et al. [255], [b]Ludlam et al. [176], [c]Middleton et al. ([209], as GRS 1915+105 is extreme and spectral modelling is inherently degenerate, we include this value for completeness: see also McClintock and Remillard [192] for additional discussion), [d]Miller and Miller [214], [e]Steiner et al. [311], [f]Middleton et al. [203]

due to a two-temperature plasma or a combination of thermal and nonthermal electron populations (e.g. Parker et al. [255]). When considering the spin value obtained
via this technique, one must therefore take into account errors resulting from the combination of detector effects and model uncertainty (e.g. the radial profile discussed in
Reynolds and Fabian [272]) and also any uncertainty in the continuum onto which the
reflection spectrum is imprinted. Naturally, as the count rate is typically extremely
high, these will dominate over statistical errors. As a final point, it is possible that
the Fe K_α line may be broadened due to scattering in the disc atmosphere which
can confuse measurements of the spin; however, as Steiner et al. [308] find in the
case of XTE J1550-564, the effect—whilst noticeable—is unlikely to dominate over
Doppler and GR effects.

At the time of writing, 17 BHBs (out of a total Galactic population of $\sim$30: [116])
have measurements for the spin from reflection fitting and these are shown in Table 3.1
(alongside those from continuum fitting). As discussed in detail by Reynolds [271],
there is general concordance with notable exceptions being 4U 1543-475 and GRO
J1655-40.

The distribution of spin values for stellar mass BHs provides insight into the natal
supernova process by which they are formed (see [153]). As remarked upon by Miller
and Miller [214], the distribution of values from the two methods is similar (although
the sample sizes do not yet allow for Gaussian-distributed statistics) with spin values
at least two orders of magnitude higher than in the case of neutron stars (where the
spin can be accurately determined from pulse periods) and demands vastly different
means of acquiring angular momentum during formation.

3.3.4   Results: AGN

Whilst the spin distribution of BHBs is considered relevant for understanding their
formation process [153], so too is it the case that the spin distribution for AGN
provides information as to the growth of SMBHs (e.g. Moderski and Sikora [224];
Madau and Quataert [178]; Volonteri et al. [333]; King and Pringle [152]; Fanidakis
et al. [87]). If the growth progressed via accretion of gas, for example driven by
minor mergers with satellite galaxies, then the infalling material will have a range of
angular momenta relative to the BH spin (see, e.g. Nayakshin et al. [236]) giving a
range of SMBH spins in the local universe. Conversely, accretion of material through
the galactic disc with a fixed direction of angular momentum is likely to produce
high spins (even if the BH is initially misaligned: [152]). As accretion is inherently
connected to the production of highly energetic outflows in the form of winds and
jets, the strength of which is potentially related to the BH spin (via the Blandford–
Znajek mechanism—see Sect. 3.3.5—and the radiative efficiency in the disc—see
Eq. 3.30), understanding how SMBHs grew and how their spin evolved is of broad
importance for our understanding of the larger scale structure of the universe due to
the interactions of outflows with the host galaxies (see the review of feedback by
Fabian [81]).

In the following sections, we discuss the campaigns to determine the spin of AGN excluding those accreting at very low (quiescent levels) such a Sgr A*, although we note that spin estimates via GRMHD simulations and SED fitting for this important SMBH have generally favoured moderate-to-high (but not maximal) spin values (e.g. Mościbrodzka et al. [227]; Dibi et al. [66]; Drappeau et al. [72]) and will be further constrained when the event horizon telescope is fully operational.

3.3.4.1 Continuum Fitting

From inspection of Eq. 3.27, it is readily apparent that the peak disc temperature scales inversely with the BH mass. For typical SMBH masses of $>10^6$ $M_\odot$ (e.g. [341]), at sub-Eddington rates, the disc will peak in the extreme UV [182] which is heavily absorbed by the ISM. Whilst substantial emission emerges at higher energies, providing the illuminating continuum for reflection, direct fitting to the disc emission to obtain the spin has not been viewed as a promising technique. In addition, for accurate spin measurements to be obtained from the continuum, it is important that the system parameters (i.e. inclination, mass and distance: see Sect. 3.3.1) are known to a high degree of precision. This can be relatively challenging; whilst the distance is known to far greater accuracy than is usually available for Galactic sources (and motivates the study of extra-galactic BHBs—[203]), the mass and inclination are harder to measure, with typical errors on the mass via reverberation of 0.5 dex.

A subclass of AGN—specifically some narrow-line Seyfert 1s (NLS1s)—appear to show a very hot disc component (e.g. Middleton et al. [210], Jin et al. [138, 139], Terashima et al. [324]) which may extend into the soft X-ray band and contribute to the ubiquitous 'soft excess' [50, 110] directly or via Comptonisation [67]. Recent progress in modelling the AGN inflow—most notably by applying a more rigourous treatment for the opacity balance (see Sect. 3.3.1)—has led to the development of the model OPTXAGNF which incorporates approximate radiative transfer, somewhat analogous to BHSPEC [67]. This model takes in the mass accretion rate which can be determined from the optical flux where the mass is reasonably well estimated and conserves energy extracted from the accretion process in creating an optically thick thermal Compton component and the higher energy tail [59]. Done et al. [68] applied this new model with a relativistic convolution kernel (KERRCONV) to the spectrum of the bright NLS1, PG 1244+026. The spectrum of this source is similar to other extreme, soft NLS1s (e.g. RE J1034+396: [204]), appearing to have high (close to the Eddington limit) mass accretion rates with a very weak tail of hard emission and a strong soft excess. Whilst high Eddington accretion rates are not generally considered to be an appropriate regime to apply standard continuum fitting methods for BHBs (see Sect. 3.3.1), analogous behaviours of the AGN disc have not been established (due to the timescales of variability). As this model conserves the energy produced via accretion (and is therefore related to the spin) in creating the regions of different optical depth, the limits on the spin imposed by this model are therefore of interest (although as with more standard models does not yet take into account energy lost in the creation of a wind or jet). In the case of PG 1244+026, Done et al.

[68] find that the spin must be low (and rule out maximal spin). Such an approach is likely to be useful for those lower mass AGN where the broadband (optical to X-ray) disc emission can be well modelled, i.e. relatively unobscured by gas and dust. Whilst these appear relatively rare (see Middleton et al. [210]), the future missions of *eROSITA* and *Athena* are expected to find many more, allowing this approach to be more widely used and thoroughly tested.

3.3.4.2   Reflection Fitting

The vast majority of spin measurements for AGN have come from fitting reflection models to their spectra. By AGN in this context, we are generally referring to Seyfert 1 AGN (including NLS1s and QSOs) as the unified model [8, 9] would suggest that these are viewed at low-to-moderate inclinations and so are not obscured by the molecular torus (which is the origin of Compton-thick AGN where Fe K_α emission is seen but originates from the torus and so is not a measure of the BH spin). The review of Reynolds [271] provides details of the approaches discussed already as well as a 'cookbook' for obtaining estimates for the BH spin from the reflection spectrum, and we direct the interested reader and practical observer here.

The Fe K_α line is known to be almost ubiquitous in AGN [235, 275] as a direct result of X-ray illumination of optically thick (not fully ionised) material and was first discovered by Tanaka et al. [321] when the bright AGN, MCG-6-30-15, was observed by *ASCA* with the strong, broad line found to require maximal spin (see also Iwasawa et al. [135]; Dabrowski [51]; Reynolds and Begelman [273] Young et al. [343]). Since the inception of the field, the use of the reflection spectrum (the whole of which is pivotal for accurate spin determination: [343]) has been widespread, finding an application in probes of higher redshift, lensed QSOs [269, 335] as well as in novel techniques to utilise the time dependence of the emission ([277]—see Sect. 3.5).

There are a number of important considerations when fitting the reflection spectrum of AGN which are not relevant in the application of this method to BHBs. The optical/UV line ratios of Seyfert AGN (e.g. Warner et al. [336]; Nagao et al. [234]) imply that the metal abundance of the gas is supersolar (e.g. Zoghbi et al. [346]; Reynolds et al. [276]) which plays an important role in the strength of the Fe K_α line; it is therefore important to allow any applied models to extend beyond solar metallicities. Unlike accretion discs around stellar mass BHs, the discs around SMBHs do not contribute in a sizeable way to the X-ray continuum (with the exception of the very brightest, lowest mass AGN: [67, 139, 210]) giving a much cleaner view of the reflection spectrum. However, the environment of the AGN is less 'clean' than in BHBs: winds are expected to be ubiquitous given the UV radiation field and resonant line opacity [263] and will not be as ionised as those from BHBs. These winds can therefore result in a distorting imprint around the Fe K_α line and possible degeneracy in the spectral fitting of AGN (e.g. Patrick et al. [256]; Brenneman et al. [34]; Middleton et al. [202]). This degeneracy can be broken in two separate ways: the first is through the use of an observable bandpass which extends to higher energies, where differences between absorption and reflection differ dramatically (as demonstrated in the case of NGC 1365 through use of *NuSTAR* data: [278]). The second means to break the spectral degeneracy is through the use of techniques which utilise the source variabliity (e.g. the cross-spectrum: Nowak and Vaughan [245]) which can effectively isolate reverberation of the primary continuum [327]. The latter technique has clearly demonstrated that the lags at high frequencies (i.e. those originating from the most compact regions of the corona) contain a signature of the broad Fe K_α line [146, 347] and Compton hump [143, 144], confirming the strong contribution of reflection to the spectrum.

A further important consideration when applying the reflection fitting method to AGN is the presence of the soft excess. As the name implies, this is an excess of flux seen below 2 keV once a power law fit to the 2–10 keV band is extrapolated backwards and has no obvious counterpart in BHB spectra. The soft excess is both smooth and is present across a large number of AGN across a range in mass, peaking at $\sim$0.5 keV in each; this raises problems for its interpretation as a continuum-only component should not produce the same peak temperature across a range of masses without some fine-tuning of the radiative process [109]. Instead, atomic transitions associated with OVII/OVIII and the Fe M UTA naturally produce features in this energy range and could be seen in absorption and emission via reflection or in an outflow. However, to produce such an excess of flux without seeing sharp features requires velocity broadening in either situation. Crummy et al. [46] and Middleton et al. [210] studied a sample of AGN, finding that both reflection and absorption can produce statistically indistinguishable fits (across the *XMM-Newton* bandpass) with the spin tending towards large values and the outflow velocity of the wind tending to being moderately relativistic. A third possibility is the combination of reflection, outflow and, in the case of the brightest, low mass AGN, some Compton component of the disc [67, 140]. Understanding the origin of the soft excess is considered to be extremely important as, in cases where the inclination of the disc is unknown or cannot be constrained in the reflection model, the spin can be driven by the need to smear the atomic features at soft energies. Once again, applying methods which utilise the time domain can assist in understanding the origin of this component and has shown that for some sources, the likely origin is in partially ionised reflection (notably 1H0707-495 [83, 346]), whilst in others it would appear that the soft excess is dominated by a continuum component associated with Compton upscattering of UV seed photons [140].

In the case where the inclination is well constrained by the model (and therefore the result does not rely on the soft excess) and the whole reflection spectrum is considered (allowing for super-solar abundances of metals), the uncertainty on the spin value once again depends on the continuum (including the effect of absorption) and the model being used, whilst at typical AGN fluxes, detector issues which can be challenging for observations of BHBs become less troublesome.

In general, SMBH spins determined via reflection (see, for example the sample studies of Walton et al. [334] and Patrick et al. [256]) show a tendency towards high spin (as remarked upon by Reynolds [271]). As discussed by Brenneman et al. [33] and Walton et al. [334], this distribution may be a result of selection effects; a higher spin results in higher radiative efficiency (see Eq. 3.30) and a brighter AGN; conversely, this may be indicating coherent rather than chaotic SMBH growth which would result in systematically high spins (see Fanidakis et al. [87] although see Nayakshin et al. [236] for counterarguments).

3.3.5 Implications: Powering of Ballistic Jets

As remarked upon in Sect. 3.2.1 of this chapter, the effect of frame-dragging will have important consequences for the transfer of energy from the BH to an orbiting test particle. Here, we briefly describe the Penrose process [258]; [18] and how it can be linked via the Blandford–Znajek mechanism to the powering of relativistic ejections.

3.3.5.1 The Penrose Effect and Blandford–Znajek Mechanism

The mechanical Penrose effect occurs as a result of 'negative energy orbits', i.e. where the energy required to send a body to infinity is larger than its rest mass energy. In the notation of GR, this is $u_t < 0$ and can be shown to only be possible for $u_\phi > 0$ and:

$$(u_\phi)^2(-g_{tt}) > \Delta \sin^2 \theta \qquad (3.39)$$

where Δ is given in Eq. 3.4. As the right-hand side of the inequality is positive, we find that this is only satisfied if $g_{tt} < 0$, i.e. within the ergosphere. It can then be shown that for a particle to have negative energy, i.e. $u_\phi > 0$, orbits need to be retrograde relative to the BH spin. The Penrose process can then be described as a body entering the ergosphere and breaking apart; one part is induced into a retrograde orbit and the other escapes to infinity after having gained energy (equal to the negative energy captured by the BH) from the rotation of the BH. The efficiency of this process, i.e. the ratio of maximum energy out to that going in, is of order 20 % (Chandrasekhar [43], see also the review of Brito et al. [35]).

Whilst the mechanism described above is unlikely to have a direct impact on the observational appearance of BHs, when coupled to magnetic fields, it may present a viable mechanism for powering collimated, relativistic outflows in the form of jets. Poloidal magnetic fields are expected to grow in the accretion flow via the MRI: [14] and propagate down to the vicinity of the BH. Where these field lines thread the ergosphere, they are forced to rotate with the matter, inducing a force on the coupled charged plasma (Lorentz force) which will lead to acceleration of material at relativistic speeds along the rotation axis of the BH in the form of jets (which are then collimated via magnetic confinement). This is a highly simplified description of the Blandford–Znajek (BZ) mechanism (or effect: [29]) where the power that can be extracted $P_{jet} \propto a^2$ as long as the spin is not large. A more accurate relationship that covers the whole range of possible spins has been derived by Tchekhovskoy and McKinney [323] to be $P_{jet} \propto (M\Omega_H)^2$ where Ω_H is the BH angular frequency which (in natural units) is given by:

$$\Omega_H = \frac{a}{2M(1 + \sqrt{1 - a^2})} \tag{3.40}$$

Measuring both the spin and jet power in a reliable way can therefore provide insights into the launching of jets by confirming the BZ effect (or an effect with a similar form) or by ruling it out.

3.3.5.2 Testing for the Blandford–Znajek Effect

As briefly mentioned in Sect. 3.3.3, BHBs launch jets at low bulk Lorentz factors ($\Gamma < 2$) when accreting at low to high rates and when accompanied by an X-ray spectrum dominated by a hard thermal tail of emission. The emission from the jet extends from low (MHz) frequencies up to optical/IR and is typically a flat power law ($S \propto E^\gamma$ where $\gamma \approx 0$) across the intervening several decades in frequency. The spectrum results from highly (ultrarelativistic) electrons spiralling around magnetic field lines and cooling via synchrotron radiation. The flat spectrum is a result of viewing the emission from spatially extended regions, with the low-frequency radio emission originating further from the launching point (i.e. where the particle Lorentz factor is lower). A break at low frequencies occurs when the lowest energy electron population becomes optically thick to their radiation (and are self-absorbed), whilst a high-frequency break occurs when the most energetic electrons are optically thin to their radiation (in reality, these breaks occur throughout the spectrum and merely convolve to give the observed spectrum). Although the spectrum evolves (notably the position of the high-frequency break moves as the electrons cool, e.g. Russell et al. [291]), as the emission is constant, the jet is termed 'steady'.

The jet changes dramatically in nature when the spectrum softens into the intermediate and then soft states, gaining a much higher bulk Lorentz factor ($\Gamma > 2$) and taking the form of discrete ejecta (e.g. Mirabel and Rodríguez [220]; Fender et al. [90]) which cool via synchrotron radiation with a high-frequency spectral break evolving with the expansion of the ejecta (see van der Laan [330]; Kellermann and Owen [149]; Hjellming [123]; Hjellming Johnston [124]). These ejections were first observed in the Galactic centre 'Annihilator', 1E 1740.7-2942 [221], and subsequently detected in the remarkable BHB, GRS 1915+105, which was the first Galactic source where superluminal jets (appearing as such due to their highly relativistic velocities and orientation close to the line-of-sight) were identified; due to their resemblance to the superluminal ejections from radio loud quasars, these sources were dubbed 'microquasars' [222]. Such discrete ejections are thought to be ubiquitous, with *all* BHBs showing (or expected to show) what are sometimes referred to as 'ballistic' jets.

Typically, the power in the steady jet is determined from the established correlation with the radio synchrotron luminosity [30, 136, 159], whilst the power in the ballistic jet has traditionally been estimated from the monochromatic radio luminosity which Steiner et al. [310] have shown to be approximately linearly correlated with the mechanical (bulk kinetic) jet power (see their Appendix). To obtain the *intrinsic* luminosity, the effect of Doppler boosting (which acts on both the flux and energy of any breaks in the synchrotron spectrum) has to be accounted for. In practice, the boosting factor ($D^{3-\gamma}$: see Eq. 3.34) is determined from the inclination (although, as discussed in Sect. 3.2.1.1, one has to careful as to which inclination—inner or outer disc—is being used) for a range of bulk Lorentz factors and assuming a typical spectral index. By correlating this deboosted jet power against the spin derived from the spectrum using the continuum fitting or reflection fitting methods, the likely impact of the BZ (or a similar) mechanism can then be tested.

Fender et al. [91] demonstrate that there is no obvious correlation between the steady jet power and the spin for a sample of BHBs, effectively ruling out the BZ effect as the dominant mechanism for the launching of the slower jet. To investigate the powering of the faster, ballistic jet, Narayan and McClintock [240] and Steiner et al. [310] selected the five Galactic BHBs which are thought to reach their Eddington limit and therefore act as 'standard candles', thereby removing any mass accretion rate-dependent effects; the resulting presence of a correlation between jet power and spin has been claimed as strong support of the BZ effect. Russell et al. [292] have since questioned the selection criteria arguing that the inclusion of other sources disagrees with the presence of a correlation; instead, the driving factor is claimed to be the mass accretion rate (with similar arguments proposed for AGN: [155, 156]). Whilst the debate is ongoing, it is abundantly clear that a larger sample of BHBs is required in order to fully evaluate the presence of a correlation. The application of these techniques to nearby extra-galactic BHBs (which can be studied reliably in the X-ray and radio: [205]) is in its infancy but is the only means by which this can be accomplished [203]. In addition, the introduction of new methods for measuring the spin (as discussed in the following sections) will provide important cross-checks for existing methods and provide increased confidence in any conclusions which rest upon its measurement.

As a final remark, Garofalo et al. [104] note that the BH spin itself may have a distorting influence on the nature of any correlation between jet power and spin through the quenching of the jet by winds which become stronger with increasing source brightness (e.g. [242]). This assumes that the winds are radiatively driven (either thermal via reprocessing or radiation pressure powered via scattering) such that a higher spin—which leads to a higher efficiency in conversion of rest mass to energy via Eq. 3.30—more readily powers winds and could in principle shut off the jet at lower mass accretion rates (under the assumption that the wind launching is not dependent on the mass accretion rate-dependent structure of the disc itself). Whilst the exact mechanism for jet quenching is not yet fully understood, certainly the coupled interaction of inflow and outflow and the spin dependence is an important consideration for understanding the evolution of accreting BHs of all masses (e.g. Kovács et al. [160]).

3.3.6   Implications: Retrograde Spins?

In a very small number of cases, only three BHBs [203, 226, 268] to date, significantly retrograde ($a_* < 0$) spin has been reported. It is possible that such sources could launch particularly powerful jets should magnetic flux be swept from the plunging region onto the BH, with the amount of flux being dependent on the size of the 'gap' between the ISCO and BH. As this is larger for retrograde spin ($R_{ISCO} = 9R_g$ for $a_* = -0.998$ (see Fig. 3.2), the magnetic flux trapped on the BH can therefore be enhanced [105, 106] although simulations incorporating the effect of magnetic field saturation [323] dispute this 'gap paradigm model' and arrive at the opposite conclusion.

Irrespective of the impact, we must ask the question, how can retrograde spin practically occur in those systems where it has been reported and how likely is it? Co-alignment of disc and BH (which occurs through the action of viscous torques transferred via the Lense–Thirring and Bardeen–Peterson effects, e.g. King et al. [151]) is expected to take at least several per cent of the binary's lifetime, and so, assuming that the measurement of retrograde spin is genuine (and not an artefact of inner disc truncation due to some as-yet-unknown process), retrograde spin is likely to be an indication of the formation process (i.e. an anisotropic supernova kick: [31]), the result of wind-fed accretion, (which can produce counteraligned inflows, e.g. GX 1+4: [41] and Cyg X-1: [303, 344]) or from the tidal capture of a star [80]. In this last case, after the BH has formed whilst in a globular cluster, it is expelled by the natal kick, with a subsequent stellar capture producing a retrograde orbit. Whether retrograde spin could form in AGN is not clear but could presumably result from minor mergers with material carrying counteraligned angular momentum accreted onto the SMBH on short timescales (as on long timescales enough material can be accreted to co-align the system). Notably in AGN, the entire sub-pc disc may be misaligned with the galaxy as a result of a recent minor merger, and indeed, there is growing evidence for this from jet launching angles and the location of the molecular torus (see Hopkins et al. [125]).

3.4   Observational Tests of Spin II—The Time Domain and Relativistic Precession Model

Traditional methods of determining the spin have proven to be highly illuminating; however, methods which do not rely solely on the time-averaged spectrum have the advantage of being able to provide a semi-independent measure of the spin and test traditional models (and our understanding of GR).

As we discussed in Sect. 3.2.1.1, the effect of relativistic frame-dragging leads to precession of orbits in the innermost regions of the accretion flow due to the Lense–Thirring effect. Such precession in turn leads to epicyclic oscillations about the vertical 'nodes' (where orbits of a test particle out of the equatorial plane meets that of the ecliptic) and precession of periastron passage. When combined with the frequency of Keplerian orbit, these three frequencies form the relativistic precession model (or RPM: [313–315]). As long as the nature of the plasma in the accretion disc is not so dense as to dampen the oscillations, in practice and under the assumption that these particle orbits can leave an imprint on the source flux we could expect each of these frequencies to leave a trace of coherent power in the light curve of accreting black hole systems (both BHBs and AGN). These could naturally be associated with the quasi-periodic oscillations (QPOs) detected as narrow peaks in the power density spectrum (PDS) and seen in BHBs at low frequencies (e.g. Wijnands et al. [338]; Casella et al. [40]; Belloni et al. [24]; Motta et al. [229]), occasionally (and only detected so far in 5 BHBs) at high frequencies (e.g. Morgan et al. [225]; Remillard and McClintock [270]; Méndez et al. [198]) and recently discovered in AGN [6, 7, 112, 204, 207]. The AGN QPOs appear to be analogous to the high-frequency QPOs (HFQPOs) of BHBs [208] and, in both sets of systems, appear in a 3:2 harmonic ratio implying a common physical origin (e.g. Dexter and Blaes [65]).

In the RPM, the two HFQPO frequencies are associated with the orbital frequency, ν_ϕ, and the periastron precession frequency, ν_{per}, which in turn is given by the difference between the orbital and radial epicyclic frequencies: $\nu_{per} = \nu_\phi - \nu_r$. The much slower, vertical (Lense–Thirring) precession would instead be associated with the low-frequency QPO (LFQPO: [130, 132]) and is given by $\nu_{lt} = \nu_\phi - \nu_\theta$ (where ν_θ is the vertical epicyclic frequency). The three QPO frequencies are connected both to one another and to the BH mass and spin through the following relations [18, 200]:

$$\nu_\phi = \pm \frac{\beta}{M} \frac{1}{r^{3/2} \pm a} \tag{3.41}$$

$$\nu_{per} = \nu_\phi \left[1 - \sqrt{1 - \frac{6}{r} \pm \frac{8a}{r^{3/2}} - \frac{3a^2}{r^2}} \right] \tag{3.42}$$

$$\nu_{lt} = \nu_\phi \left[1 - \sqrt{1 \mp \frac{4a}{r^{3/2}} + \frac{3a^2}{r^2}} \right] \tag{3.43}$$

where $\beta = c^3/(2\pi GM_\odot)$, r is the radius in units of R_g and M is in units of solar mass. Where $\pm$ or $\mp$ are given, the top sign refers to a treatment where the spin is prograde and the bottom sign to where the spin is retrograde.

The above set of equations lead to a set of simultaneous equations and in turn to the following formula for the spin (see Ingram and Motta [131] for a derivation):

$$a = \pm \frac{r^{3/2}}{4} \left[\Lambda + \Phi - 2 + \frac{6}{r} \right] \tag{3.44}$$

where:

$$\Phi = \left(1 - \frac{\nu_{per}}{\nu_\phi} \right)^2 \tag{3.45}$$

and

$$\Lambda = \left(1 - \frac{\nu_{per}}{\nu_{lt}}\right)^2 \tag{3.46}$$

Although the solutions to the equations of the RPM do not immediately tell us whether the spin is prograde or retrograde, Ingram and Motta [131] point out that this can be identified from the highest frequency reached by the LFQPO (which if it extends to within the ISCO for $a = -|a|$, then it implies prograde spin).

Critically, the application of the RPM in determining the spin relies on the *simultaneous* presence of LF and HFQPOs in the light curve (as their frequencies are indicative of the radius at which they are generated), although it is not vital that all three be present as Ingram and Motta [131] demonstrate semi-analytically. Recently, the RPM method has been applied to two BHBs: GRO J1655-40 [228], the only BHB to date where all three QPOs have been detected simultaneously [229] and XTE J1550-654 [231] where only the LFQPO and one of the HFQPOs have been detected simultaneously. The resulting spin for XTE J1550-654 was found to be consistent with estimates from the reflection fitting and continuum method [308], whereas the value obtained for GRO J1655-40 via the RPM is inconsistent with those obtained via spectral means [213, 267, 300]. The reason for the discrepancy in the spin for the latter source may be due to the misalignment of the BH spin and orbital axis by >15° [114] which are likely to be closely aligned in XTE J1550-654 [309]. Thus, whilst the RPM is insensitive to the inclination (as the frequencies are independent of the inclination unlike the measured strengths of the QPO: [230]), this could present problems for measurements of the spin via spectroscopic methods (see the discussions in Sect. 3.3).

Finally, it should be pointed out that mechanisms to explain the origins of the QPOs besides the RPM have also been proposed (e.g. Esin et al. [79]; Tagger and Pellat [320]), with little *direct* observational progress made in distinguishing the correct interpretation. However, a key discriminator for the origin of the LFQPO as Lense–Thirring precession is the QPO phase-resolved emission (see Sect. 3.3.2). As described in Ingram and Done [130], the geometrically thin, optically thick disc truncates (possibly due to disc evaporation, e.g. Meyer and Meyer-Hofmeister [201]; Liu et al. [175]) and the inner, lower density region (in this picture, the location of the Compton upscattering of seed disc photons) precesses as a solid body as a result of the Lense–Thirring effect (see Ingram et al. [133]). As the inner region precesses, various disc azimuths are subjected to changing illumination; in turn, this leads to changes in the observed reflected emission (Fe lines and Compton hump) as a result of the various Doppler shifts and boosting. Should these predicted changes be observed by long observations with present instruments or using high throughput, high-time-resolution instruments such as the LAXPC onboard the recently launched *ASTROSAT* (see Ingram et al. [134] for details of the arithmetic approaches and a possible detection of modulation already seen in *RXTE* data), it will likely represent a 'smoking gun' for the RPM (and may likewise rule out such an origin if the predicted variations are not observed).

3.5 Observational Tests of Spin III—The Energy–Time Domain

As opposed to the previously described methods which use either the energy *or* time domain, an approach which combines the two, promises to provide the largest lever arm for estimating the BH spin. One such method is 'Doppler tomography' and relies upon a changing view of the regions of the accretion disc due to an eclipse by an orbiting body. This technique was first applied to the study of white dwarfs in mapping the accretion disc via emission lines (e.g. Marsh and Horne [186]), and a small number of authors have since developed it as a tool for the study of AGN spin and the effects of GR.

In the majority of AGN (with masses typically $> 10^6 \, M_\odot$: [341]), the disc is out of the X-ray bandpass (although the hottest tail of the disc and/or a Compton upscattered component may enter at soft energies for the lowest mass and highest mass accretion rate sources: [210]; Jin et al. [138]; [67], see the discussion in Sect. 3.3.1), whilst the primary Compton scattered emission and its reflected component dominate the emission (e.g. Fabian [83]). Should an orbiting body pass across our line-of-sight, an eclipse results and leads to changes in the spectrum as a function of time that allows a test of the nature of the inner regions, e.g. the radial temperature dependence of the corona and BH spin (see McKernan and Yaqoob [197]).

Variability due to obscuration by cold material is relatively common on long timescales in AGN (see [279]) and on shorter timescales by Compton-thin material [25, 76, 264, 277, 280]. However, obscuration by Compton-thick (i.e. $\tau > 1 \rightarrow n_H \sigma_T > 1 \rightarrow n_H > 10^{24} \, \text{cm}^{-2}$) material on observable (< 100 s of ks) timescales— which leads to the simplest form of eclipse—is relatively rare, although at least one such event has been observed in NGC 1365 [282].

Using the model of Dovčiak et al. [71], which calculates the line emission from different parts of the disc separately, Risaliti et al. [281] simulate the effects of an eclipse under the assumption that the obscuring source is a Compton-thick broad-line region cloud and completely covers the source. The result is a profound shift in the shape of the Fe K_α line as a function of time as the approaching blue-shifted side of the disc is covered followed by the retreating red-shifted side (i.e. a situation where the cloud is co-rotating with the disc). As Risaliti et al. [281] point out, the major effect is not a shift in the line profile but in flux due to Doppler boosting. As a corollary, the high-energy continuum emission, i.e. the Compton hump, will also rise and fall in flux, correlated with changes in the emission line (Fig. 3.1 of [281]). Risaliti et al. [281] point out that this would constitute a case of a 'perfect' eclipse as the obscuration is complete and the eclipse assumed to have sharp, linear edges. The authors perform simulations which show that present observatories have the required throughput to detect predicted changes for Compton-thick eclipses and can provide independent confirmation of relativistic effects in shaping the Fe K_α line. Although the more common eclipses by Compton-thin material (typically a few $10^{23} \, \text{cm}^{-2}$: [180, 282]) have less of an impact on the reflection spectrum, Risaliti et al. [281]

show that future observatories (for instance ESA's *Athena*) will still be able to detect changes associated with the passage of such material.

Although not modelled explicitly, the change in Doppler boosting from either side of the disc is dependent on both the inclination to the source and rotational velocity (see Eq. 3.34). The latter is of course dependent on the location of the ISCO and therefore the spin. Doppler tomography therefore not only provides a means to independently verify the origin of the emission in relativistic material but also a measurement of the spin. There are of course qualifiers and caveats to this approach, and Risaliti et al. [281] point out that variations in the intrinsic emission can lead to a distorting effect as can an eclipse that covers the illuminating source in a different manner to the reflector.

An analogous situation to that described by Risaliti et al. [281] can be applied to the disc emission directly when eclipses take place in AGN of lower mass. Such low mass AGN are preferentially detected at the highest mass accretion rates [115], where, analogous to the spectra of BHBs [138, 139], the spectrum is dominated by the disc with a weak, flat power-law tail of emission to high energies. In such sources, reflection features are therefore expected to be weak, restricting the means by which the spin can be measured. However, at the apparent high accretion rates associated with low-mass AGN, powerful winds are expected to be driven from the disc, which itself may have grown in scale height (e.g. Shakura and Sunyaev [301]; King [154]; Poutanen et al. [260]). As such winds are likely to rotate in an approximately Keplerian manner (with deviations from this expected to scale as $(H/R)^2$), any inhomogeneities due to radiative hydrodynamic instabilities in the surface of the wind material [263] will lead to gaps through to the inner regions which also rotate. Should our sight line to the source intercept one of these gaps, we can obtain a view of the approaching, blueshifted side of the disc and then the retreating redshifted side as the gap orbits (see Fig. 3.5). As with the model presented by Risaliti et al. [277], this form of Doppler tomography is highly sensitive to the spin and inclination and can therefore be used to provide independent constraints. Notably, unlike studies of the reflection, the disc emission is expected to be stable over the timescales of an observation and so is not likely to be affected by intrinsic variability that can have a distorting influence.

Using the ray-tracing code GEOKERR [62], Middleton and Ingram [211] create a model to describe the orbit of a gap in a Compton-thick wind and apply it to the case of the low-mass AGN RX J1301.9+2747 [318] which shows long-lived flaring behaviour (Dewangan et al. [61]) inconsistent with the usual origin of rapid variability in AGN (i.e. viscous—see [328]—and/or thermal). The authors instead argue that the variability is due to gaps in the Compton-thick wind crossing our line-of-sight (Fig. 3.5). From fitting the model across multiple phases simultaneously, the authors find the spin to be very low irrespective of several caveats (e.g. errors on the mass and temperature profile in the disc) although once again important assumptions remain including the unknown structure of the disc seen through the gaps (the ray-tracing assumes a Novikov–Thorne disc). Importantly, the combination of the time and energy domains leads to stronger constraints on the spin than can be obtained from traditional methods, demonstrating the power of Doppler tomography as a method to probe AGN accretion and the region of strong gravity.

3.6 Concluding Remarks and Future Approaches

In this chapter we have discussed the core theory that is useful for an appreciation of the role BH spin plays, notably the effect of precession and the Penrose process due to frame-dragging and the changing position of the ISCO. As a result of the latter's effect on the emergent radiation (be it direct or reflected), the community has been provided with a means to measure the BH spin in both AGN and BHBs.

Campaigns over the last 10 years have started to allow the spin distributions of BHs to be probed, allowing progress to be made in understanding their formation. However, many questions remain open and as yet unanswered: What role does the spin play in the launching of ballistic jets? Is there a bias in the spin measurements of AGN or is the spin genuinely high in most local Seyferts? Is retrograde spin common or vanishingly rare? How reliable are our present set of techniques?

The first three questions can only be addressed by expanding our sample of sources for which we have accurate spin measurements. This will no doubt be possible in the forthcoming years when new, highly sensitive X-ray satellites including *Athena* and *eROSITA* (and potentially *LOFT* or a descendent) become available. These will provide the deepest views of the X-ray universe, providing access to not only the spins of local sources but also, in the case of *Athena*, the cosmic evolution of the AGN spin distribution (which in turn probes the growth mechanism of SMBHs: [87, 333]). The photon-rich spectra that *Athena* will obtain will not only provide high-precision spin measurements but potentially even test for deviations from the Kerr metric (e.g. Jiang et al. [137]).

As we have discussed in Sect. 3.3.5, to rigorously probe the BZ effect (in BHBs—the analogy to AGN jets is still not clear) requires a much larger sample of sources with reliable measures of both BH spin and jet power—for this, we must look to nearby galaxies. Such an approach has been proven to be feasible with current instrumentation [203] and in future will benefit from the introduction of both high-throughout X-ray instruments and the next generation of radio telescopes (*SKA* and pathfinders), for which the discovery of radio transients is a core aim (e.g. the ThunderKAT campaign).

Finally, to test the reliability of our techniques requires the use of the time domain in an independent (RPM) or complimentary (Doppler tomography) fashion and looking to the future, the use of X-ray polarimetry and gravitational wave interferometry. Expanding briefly on the latter two techniques, as explained in Schnittman and Krolik [297], the effect of returning radiation from the accretion disc leads to scattering which is not appreciable at low energies (i.e. further out in the disc) and leads to horizontal polarisation [42], whilst at higher energies the increased scatter to the observer results in vertical polarisation [5]. The amount of the latter is dependent on the position of the ISCO and therefore the spin. This is expected to be an extremely powerful technique and is the focus of a number of proposed (e.g. *PRAXyS* and *X-Calibur*) and accepted (*XIPE*) missions. The impact of gravitational wave interferometry on the field of BH spin measurements has been discussed in detail in the recent review by Miller and Miller [214] to which we point the interested reader; in essence, should a BH–NS or BH–BH binary be found and if the BH spin is high and misaligned with the orbital axis (due to LT precession: see Sect. 3.2.1), then there can be a considerable impact on the gravitational waveform.

In conclusion, the future looks extremely bright for the field of BH spin determination and, in years to come, will allow the most detailed understanding of the most extreme objects in the universe.

Acknowledgments The author gratefully acknowledges the assistance of Chris Reynolds, Javier Garcia and Jack Steiner in proofreading and offering valuable suggestions.

References

1. M.A. Abramowicz, P.C. Fragile, LRR **16**, 1 (2013)
2. M.A. Abramowicz, B. Czerny, J.P. Lasota, E. Szuszkiewicz, ApJ **332**, 646 (1988)
3. N. Afshordi, B. Paczyński, ApJ **592**, 354 (2003)
4. E. Agol, PhDT (1997)
5. E. Agol, J.H. Krolik, ApJ **528**, 161 (2000)
6. W.N. Alston, M.L. Parker, J. Markevičiūtė, A.C. Fabian, M. Middleton, A. Lohfink, E. Kara, C. Pinto, MNRAS **449**, 467 (2015)
7. W.N. Alston, J. Markevičiūtė, E. Kara, A.C. Fabian, M. Middleton, MNRAS **445**, L16 (2014)
8. R.R.J. Antonucci, J.S. Miller, ApJ **297**, 621 (1985)
9. R. Antonucci, ARA&A **31**, 473 (1993)
10. P.J. Armitage, P. Natarajan, ApJ **525**, 909 (1999)
11. P.J. Armitage, C.S. Reynolds, J. Chiang, ApJ **548**, 868 (2001)
12. K.A. Arnaud, ASPC **101**, 17 (1996)
13. S.A. Balbus, MNRAS **423**, L50 (2012)
14. S.A. Balbus, J.F. Hawley, ApJ **376**, 214 (1991)
15. S.A. Balbus, J.F. Hawley, RvMP **70**, 1 (1998)
16. D.R. Ballantyne, N.J. Turner, O.M. Blaes, ApJ **603**, 436 (2004)
17. J.M. Bardeen, B. Carter, S.W. Hawking, CMaPh **31**, 161 (1973)
18. J.M. Bardeen, W.H. Press, S.A. Teukolsky, ApJ **178**, 347 (1972)
19. J.M. Bardeen, J.A. Petterson, ApJ **195**, L65 (1975)
20. P. Barr, N.E. White, C.G. Page, MNRAS **216**, 65P (1985)
21. M.C. Begelman, C.F. McKee, G.A. Shields, ApJ **271**, 70 (1983)
22. T. Belloni, M. Méndez, A.R. King, M. van der Klis, J. van Paradijs, ApJ **479**, L145 (1997)
23. T.M. Belloni, L. Stella, SSRv **183**, 43 (2014)
24. T.M. Belloni, S.E. Motta, T. Muñoz-Darias, BASI **39**, 409 (2011)
25. S. Bianchi, E. Piconcelli, M. Chiaberge, E.J. Bailón, G. Matt, F. Fiore, ApJ **695**, 781 (2009)
26. G.S. Bisnovatyi-Kogan, B.V. Komberg, SvA **18**, 217 (1974)
27. O. Blaes, S. Hirose, J.H. Krolik, ApJ **664**, 1057 (2007)
28. R.D. Blandford, M.C. Begelman, MNRAS **303**, L1 (1999)
29. R.D. Blandford, R.L. Znajek, MNRAS **179**, 433 (1977)
30. R.D. Blandford, A. Königl, ApJ **232**, 34 (1979)
31. N. Brandt, P. Podsiadlowski, MNRAS **274**, 461 (1995)

32. L.W. Brenneman, C.S. Reynolds, ApJ **652**, 1028 (2006)
33. L.W. Brenneman et al., ApJ **736**, 103 (2011)
34. L.W. Brenneman, M. Elvis, Y. Krongold, Y. Liu, S. Mathur, ApJ **744**, 13 (2012)
35. R. Brito, V. Cardoso, P. Pani, *Lecture Notes in Physics* (Springer, Berlin, 2015), p. 906
36. L.M. Burko, A. Ori, AnIPS, 13 (1997)
37. E.M. Cackett, A. Zoghbi, C. Reynolds, A.C. Fabian, E. Kara, P. Uttley, D.R. Wilkins, MNRAS **438**, 2980 (2014)
38. A. Čadež, C. Fanton, M. Calvani, New A **3**, 647 (1998)
39. S. Campana, L. Stella, MNRAS **264**, 395 (1993)
40. P. Casella, T. Belloni, L. Stella, ApJ **629**, 403 (2005)
41. D. Chakrabarty et al., ApJ **481**, L101 (1997)
42. S. Chandrasekhar, ratr.book (1960)
43. S. Chandrasekhar, mtbh.book (1983)
44. E. Churazov, M. Gilfanov, M. Revnivtsev, MNRAS **321**, 759 (2001)
45. S. Corbel, E. Koerding, P. Kaaret, MNRAS **389**, 1697 (2008)
46. J. Crummy, A.C. Fabian, L. Gallo, R.R. Ross, MNRAS **365**, 1067 (2006)
47. C.T. Cunningham, J.M. Bardeen, ApJ **183**, 237 (1973)
48. C.T. Cunningham, ApJ **202**, 788 (1975)
49. C. Cunningham, ApJ **208**, 534 (1976)
50. B. Czerny, M. Nikołajuk, A. Różańska, A.-M. Dumont, Z. Loska, P.T. Zycki, A&A **412**, 317 (2003)
51. Y. Dabrowski, A.C. Fabian, K. Iwasawa, A.N. Lasenby, C.S. Reynolds, MNRAS **288**, L11 (1997)
52. X. Dai, C.S. Kochanek, G. Chartas, S. Kozłowski, C.W. Morgan, G. Garmire, E. Agol, ApJ **709**, 278 (2010)
53. T. Dauser, J. Garcia, J. Wilms, M. Böck, L.W. Brenneman, M. Falanga, K. Fukumura, C.S. Reynolds, MNRAS **430**, 1694 (2013)
54. T. Dauser, J. Wilms, C.S. Reynolds, L.W. Brenneman, MNRAS **409**, 1534 (2010)
55. T. Dauser, J. García, M.L. Parker, A.C. Fabian, J. Wilms, MNRAS **444**, L100 (2014)
56. S.W. Davis, C. Done, O.M. Blaes, ApJ **647**, 525 (2006)
57. S.W. Davis, O.M. Blaes, I. Hubeny, N.J. Turner, ApJ **621**, 372 (2005)
58. S.W. Davis, I. Hubeny, ApJS **164**, 530 (2006)
59. S.W. Davis, A. Laor, ApJ **728**, 98 (2011)
60. B. De Marco, G. Ponti, M. Cappi, M. Dadina, P. Uttley, E.M. Cackett, A.C. Fabian, G. Miniutti, MNRAS **431**, 2441 (2013)
61. G.C. Dewangan, K.P. Singh, Y.D. Mayya, G.C. Anupama, MNRAS **318**, 309 (2000)
62. J. Dexter, E. Agol, ApJ **696**, 1616 (2009)
63. J. Dexter, E. Agol, ApJ **727**, L24 (2011)
64. J. Dexter, E. Quataert, MNRAS **426**, L71 (2012)
65. J. Dexter, O. Blaes, MNRAS **438**, 3352 (2014)
66. S. Dibi, S. Drappeau, P.C. Fragile, S. Markoff, J. Dexter, MNRAS **426**, 1928 (2012)
67. C. Done, S.W. Davis, C. Jin, O. Blaes, M. Ward, MNRAS **420**, 1848 (2012)
68. C. Done, C. Jin, M. Middleton, M. Ward, MNRAS **434**, 1955 (2013)
69. C. Done, M. Gierliński, A. Kubota, A&ARv **15**, 1 (2007)
70. T. Dotani et al., ApJ **485**, L87 (1997)
71. M. Dovčiak, V. Karas, T. Yaqoob, ApJS **153**, 205 (2004)
72. S. Drappeau, S. Dibi, J. Dexter, S. Markoff, P.C. Fragile, MNRAS **431**, 2872 (2013)
73. R.J.H. Dunn, R.P. Fender, E.G. Körding, T. Belloni, C. Cabanac, MNRAS **403**, 61 (2010)
74. K. Ebisawa, K. Mitsuda, T. Hanawa, ApJ **367**, 213 (1991)
75. T. Ebisuzaki, D. Sugimoto, T. Hanawa, PASJ **36**, 551 (1984)
76. M. Elvis, G. Risaliti, F. Nicastro, J.M. Miller, F. Fiore, S. Puccetti, ApJ **615**, L25 (2004)
77. D. Emmanoulopoulos, I.M. McHardy, I.E. Papadakis, MNRAS **416**, L94 (2011)
78. D. Emmanoulopoulos, I.E. Papadakis, M. Dovčiak, I.M. McHardy, MNRAS **439**, 3931 (2014)
79. A.A. Esin, J.E. McClintock, R. Narayan, ApJ **489**, 865 (1997)

80. A.C. Fabian, J.E. Pringle, M.J. Rees, MNRAS **172**, 15P (1975)
81. A.C. Fabian, ARA&A **50**, 455 (2012)
82. A.C. Fabian, M.J. Rees, L. Stella, N.E. White, MNRAS **238**, 729 (1989)
83. A.C. Fabian et al., Nature **459**, 540 (2009)
84. A.C. Fabian, R.R. Ross, SSRv **157**, 167 (2010)
85. A.C. Fabian, K. Iwasawa, C.S. Reynolds, A.J. Young, PASP **112**, 1145 (2000)
86. C. Fanton, M. Calvani, F. de Felice, A. Cadez, PASJ **49**, 159 (1997)
87. N. Fanidakis, C.M. Baugh, A.J. Benson, R.G. Bower, S. Cole, C. Done, C.S. Frenk, MNRAS **410**, 53 (2011)
88. R.P. Fender, E. Gallo, P.G. Jonker, MNRAS **343**, L99 (2003)
89. R.P. Fender, T.M. Belloni, E. Gallo, MNRAS **355**, 1105 (2004)
90. R.P. Fender, J. Homan, T.M. Belloni, MNRAS **396**, 1370 (2009)
91. R.P. Fender, E. Gallo, D. Russell, MNRAS **406**, 1425 (2010)
92. T. Fragos et al., ApJ **683**, 346 (2008)
93. T. Fragos, J.E. McClintock, ApJ **800**, 17 (2015)
94. J. Frank, A.R. King, D.J. Raine, apa..book (1985)
95. E. Gallo et al., MNRAS **445**, 290 (2014)
96. C.F. Gammie, ApJ **522**, L57 (1999)
97. C.F. Gammie, S.L. Shapiro, J.C. McKinney, ApJ **602**, 312 (2004)
98. P. Gandhi et al., MNRAS **407**, 2166 (2010)
99. J. García, T.R. Kallman, ApJ **718**, 695 (2010)
100. J. García, T. Dauser, C.S. Reynolds, T.R. Kallman, J.E. McClintock, J. Wilms, W. Eikmann, ApJ **768**, 146 (2013)
101. J. García, T.R. Kallman, R.F. Mushotzky, ApJ **731**, 131 (2011)
102. J. García et al., ApJ **782**, 76 (2014)
103. J.A. García, T. Dauser, J.F. Steiner, J.E. McClintock, M.L. Keck, J. Wilms, ApJ **808**, L37 (2015)
104. D. Garofalo, M.I. Kim, D.J. Christian, MNRAS **442**, 3097 (2014)
105. D. Garofalo, D.A. Evans, R.M. Sambruna, MNRAS **406**, 975 (2010)
106. D. Garofalo, ApJ **699**, 400 (2009)
107. I.M. George, A.C. Fabian, MNRAS **249**, 352 (1991)
108. G. Ghisellini, F. Haardt, G. Matt, MNRAS **267**, 743 (1994)
109. M. Gierliński, C. Done, MNRAS **347**, 885 (2004)
110. M. Gierliński, C. Done, MNRAS **349**, L7 (2004)
111. M. Gierliński, A.A. Zdziarski, J. Poutanen, P.S. Coppi, K. Ebisawa, W.N. Johnson, MNRAS **309**, 496 (1999)
112. M. Gierliński, M. Middleton, M. Ward, C. Done, Nature **455**, 369 (2008)
113. L. Gou et al., ApJ **701**, 1076 (2009)
114. J. Greene, C.D. Bailyn, J.A. Orosz, ApJ **554**, 1290 (2001)
115. J.E. Greene, L.C. Ho, ApJ **610**, 722 (2004)
116. H.-J. Grimm, M. Gilfanov, R. Sunyaev, A&A **391**, 923 (2002)
117. P.W. Guilbert, M.J. Rees, MNRAS **233**, 475 (1988)
118. F. Haardt, L. Maraschi, G. Ghisellini, ApJ **432**, L95 (1994)
119. J.F. Hawley, J.H. Krolik, ApJ **566**, 164 (2002)
120. A. Hirano, S. Kitamoto, T.T. Yamada, S. Mineshige, J. Fukue, ApJ **446**, 350 (1995)
121. S. Hirose, O. Blaes, J.H. Krolik, ApJ **704**, 781 (2009)
122. R.M. Hjellming, M.P. Rupen, Nature **375**, 464 (1995)
123. R.M. Hjellming, gera.book, 381 (1988)
124. R.M. Hjellming, K.J. Johnston, ApJ **328**, 600 (1988)
125. P.F. Hopkins, L. Hernquist, C.C. Hayward, D. Narayanan, MNRAS **425**, 1121 (2012)
126. J.C. Houck, L.A. Denicola, ASPC **216**, 591 (2000)
127. I. Hubeny, O. Blaes, J.H. Krolik, E. Agol, ApJ **559**, 680 (2001)
128. I. Hubeny, E. Agol, O. Blaes, J.H. Krolik, ApJ **533**, 710 (2000)
129. I. Hubeny, T. Lanz, ApJ **439**, 875 (1995)

130. A. Ingram, C. Done, MNRAS **427**, 934 (2012)
131. A. Ingram, S. Motta, MNRAS **444**, 2065 (2014)
132. A. Ingram, C. Done, MNRAS **415**, 2323 (2011)
133. A. Ingram, C. Done, P.C. Fragile, MNRAS **397**, L101 (2009)
134. A. Ingram, M. van der Klis, MNRAS **446**, 3516 (2015)
135. K. Iwasawa et al., MNRAS **282**, 1038 (1996)
136. O. Jamil, R.P. Fender, C.R. Kaiser, MNRAS **401**, 394 (2010)
137. J. Jiang, C. Bambi, J.F. Steiner, JCAP **5**, 025 (2015)
138. C. Jin, M. Ward, C. Done, J. Gelbord, MNRAS **420**, 1825 (2012)
139. C. Jin, M. Ward, C. Done, MNRAS **425**, 907 (2012)
140. C. Jin, C. Done, M. Middleton, M. Ward, MNRAS **436**, 3173 (2013)
141. P.G. Jonker, G. Nelemans, MNRAS **354**, 355 (2004)
142. T. Kallman, M. Bautista, ApJS **133**, 221 (2001)
143. E. Kara et al., MNRAS **449**, 234 (2015)
144. E. Kara et al., MNRAS **446**, 737 (2015)
145. E. Kara, E.M. Cackett, A.C. Fabian, C. Reynolds, P. Uttley, MNRAS **439**, L26 (2014)
146. E. Kara, A.C. Fabian, E.M. Cackett, P. Uttley, D.R. Wilkins, A. Zoghbi, MNRAS **434**, 1129 (2013)
147. E. Kara, A.C. Fabian, E.M. Cackett, G. Miniutti, P. Uttley, MNRAS **430**, 1408 (2013)
148. V. Karas, O. Kopáček, D. Kunneriath, CQGra **29**, 035010 (2012)
149. K.I. Kellermann, F.N. Owen, gera.book, 563 (1988)
150. R.P. Kerr, mgm..conf, 9 (2008)
151. A.R. King, S.H. Lubow, G.I. Ogilvie, J.E. Pringle, MNRAS **363**, 49 (2005)
152. A.R. King, J.E. Pringle, J.A. Hofmann, MNRAS **385**, 1621 (2008)
153. A.R. King, U. Kolb, MNRAS **305**, 654 (1999)
154. A.R. King, M.B. Davies, M.J. Ward, G. Fabbiano, M. Elvis, ApJ **552**, L109 (2001)
155. A.L. King, J.M. Miller, M. Bietenholz, K. Gültekin, M. Reynolds, A. Mioduszewski, M. Rupen, N. Bartel, ApJ **799**, L8 (2015)
156. A.L. King, J.M. Miller, K. Gültekin, D.J. Walton, A.C. Fabian, C.S. Reynolds, K. Nandra, ApJ **771**, 84 (2013)
157. Y.-K. Ko, T.R. Kallman, ApJ **431**, 273 (1994)
158. M. Kolehmainen, C. Done, M. Díaz, Trigo. MNRAS **437**, 316 (2014)
159. E.G. Körding, R.P. Fender, S. Migliari, MNRAS **369**, 1451 (2006)
160. Z. Kovács, L. Gergely, P.L. Biermann, MNRAS **416**, 991 (2011)
161. J.H. Krolik, ApJ **515**, L73 (1999)
162. J.H. Krolik, C.F. McKee, C.B. Tarter, ApJ **249**, 422 (1981)
163. A. Kubota, K. Ebisawa, K. Makishima, K. Nakazawa, ApJ **631**, 1062 (2005)
164. A. Kubota, K. Makishima, ApJ **601**, 428 (2004)
165. A. Kubota, C. Done, MNRAS **353**, 980 (2004)
166. A.K. Kulkarni et al., MNRAS **414**, 1183 (2011)
167. A. Laor, ApJ **376**, 90 (1991)
168. J.-P. Lasota, NewAR **45**, 449 (2001)
169. J. Lense, H. Thirring, Phys. Z. **19**, 156 (1918)
170. L.-X. Li, Phys. Rev. D **67**, 044007 (2003)
171. L.-X. Li, E.R. Zimmerman, R. Narayan, J.E. McClintock, ApJS **157**, 335 (2005)
172. E.P.T. Liang, R.H. Price, ApJ **218**, 247 (1977)
173. A.P. Lightman, D.M. Eardley, ApJ **187**, L1 (1974)
174. A.P. Lightman, T.R. White, ApJ **335**, 57 (1988)
175. B.F. Liu, W. Yuan, F. Meyer, E. Meyer-Hofmeister, G.Z. Xie, ApJ **527**, L17 (1999)
176. R.M. Ludlam, J.M. Miller, E.M. Cackett, (2015). arXiv:1505.05449
177. T.J. Maccarone, MNRAS **336**, 1371 (2002)
178. P. Madau, E. Quataert, ApJ **606**, L17 (2004)
179. P. Magdziarz, A.A. Zdziarski, MNRAS **273**, 837 (1995)
180. R. Maiolino et al., A&A **517**, A47 (2010)

181. K. Makishima, Y. Maejima, K. Mitsuda, H.V. Bradt, R.A. Remillard, I.R. Tuohy, R. Hoshi, M. Nakagawa, ApJ **308**, 635 (1986)
182. M.A. Malkan, ApJ **268**, 582 (1983)
183. R.G. Martin, R.C. Reis, J.E. Pringle, MNRAS **391**, L15 (2008)
184. S. Markoff, M.A. Nowak, ApJ **609**, 972 (2004)
185. D. Marković, F.K. Lamb, ApJ **507**, 316 (1998)
186. T.R. Marsh, K. Horne, MNRAS **235**, 269 (1988)
187. R.G. Martin, C.A. Tout, J.E. Pringle, MNRAS **387**, 188 (2008)
188. A. Martocchia, V. Karas, G. Matt, MNRAS **312**, 817 (2000)
189. G. Matt, A.C. Fabian, R.R. Ross, MNRAS **278**, 1111 (1996)
190. G. Matt, A.C. Fabian, R.R. Ross, MNRAS **262**, 179 (1993)
191. G. Matt, G.C. Perola, MNRAS **259**, 433 (1992)
192. J.E. McClintock, R.A. Remillard, csxs.book, 157 (2006)
193. J.E. McClintock, R. Shafee, R. Narayan, R.A. Remillard, S.W. Davis, L.-X. Li, ApJ **652**, 518 (2006)
194. J.E. McClintock, R. Narayan, J.F. Steiner, SSRv **183**, 295 (2014)
195. I.M. McHardy, E. Koerding, C. Knigge, P. Uttley, R.P. Fender, Nature **444**, 730 (2006)
196. I.M. McHardy, P. Arévalo, P. Uttley, I.E. Papadakis, D.P. Summons, W. Brinkmann, M.J. Page, MNRAS **382**, 985 (2007)
197. B. McKernan, T. Yaqoob, ApJ **501**, L29 (1998)
198. M. Méndez, D. Altamirano, T. Belloni, A. Sanna, MNRAS **435**, 2132 (2013)
199. A. Merloni, A.C. Fabian, R.R. Ross, MNRAS **313**, 193 (2000)
200. A. Merloni, M. Vietri, L. Stella, D. Bini, MNRAS **304**, 155 (1999)
201. F. Meyer, E. Meyer-Hofmeister, A&A **288**, 175 (1994)
202. M.J. Middleton, M.L. Parker, C.S. Reynolds, A.C. Fabian, A.M. Lohfink, MNRAS **457**, 1568 (2016)
203. M.J. Middleton, J.C.A. Miller-Jones, R.P. Fender, MNRAS **439**, 1740 (2014)
204. M. Middleton, C. Done, M. Ward, M. Gierliński, N. Schurch, MNRAS **394**, 250 (2009)
205. M.J. Middleton et al., Nature **493**, 187 (2013)
206. M.J. Middleton, A.D. Sutton, T.P. Roberts, F.E. Jackson, C. Done, MNRAS **420**, 2969 (2012)
207. M. Middleton, P. Uttley, C. Done, MNRAS **417**, 250 (2011)
208. M. Middleton, C. Done, MNRAS **403**, 9 (2010)
209. M. Middleton, C. Done, M. Gierliński, S.W. Davis, MNRAS **373**, 1004 (2006)
210. M. Middleton, C. Done, M. Gierliński, MNRAS **381**, 1426 (2007)
211. M.J. Middleton, A.R. Ingram, MNRAS **446**, 1312 (2015)
212. J.M. Miller et al., ApJ **724**, 1441 (2010)
213. J.M. Miller, C.S. Reynolds, A.C. Fabian, G. Miniutti, L.C. Gallo, ApJ **697**, 900 (2009)
214. M.C. Miller, J.M. Miller, Phys. Rep. **548**, 1 (2015)
215. J.C.A. Miller-Jones, PASA **31**, e016 (2014)
216. S. Mineshige, A. Hirano, S. Kitamoto, T.T. Yamada, J. Fukue, ApJ **426**, 308 (1994)
217. G. Miniutti, A.C. Fabian, R. Goyder, A.N. Lasenby, MNRAS **344**, L22 (2003)
218. G. Miniutti, A.C. Fabian, MNRAS **349**, 1435 (2004)
219. I.F. Mirabel, R. Mignani, I. Rodrigues, J.A. Combi, L.F. Rodríguez, F. Guglielmetti, A&A **395**, 595 (2002)
220. I.F. Mirabel, L.F. Rodríguez, ARA&A **37**, 409 (1999)
221. I.F. Mirabel, L.F. Rodriguez, B. Cordier, J. Paul, F. Lebrun, Nature **358**, 215 (1992)
222. I.F. Mirabel, L.F. Rodríguez, Nature **371**, 46 (1994)
223. K. Mitsuda et al., PASJ **36**, 741 (1984)
224. R. Moderski, M. Sikora, MNRAS **283**, 854 (1996)
225. E.H. Morgan, R.A. Remillard, J. Greiner, ApJ **482**, 993 (1997)
226. W.R. Morningstar, J.M. Miller, R.C. Reis, K. Ebisawa, ApJ **784**, L18 (2014)
227. M. Mościbrodzka, C.F. Gammie, J.C. Dolence, H. Shiokawa, P.K. Leung, ApJ **706**, 497 (2009)
228. S.E. Motta, T.M. Belloni, L. Stella, T. Muñoz-Darias, R. Fender, MNRAS **437**, 2554 (2014)

229. S. Motta, J. Homan, T. Muñoz, Darias, P. Casella, T.M. Belloni, B. Hiemstra, M. Méndez. MNRAS **427**, 595 (2012)
230. S.E. Motta, P. Casella, M. Henze, T. Muñoz-Darias, A. Sanna, R. Fender, T. Belloni, MNRAS **447**, 2059 (2015)
231. S.E. Motta, T. Muñoz-Darias, A. Sanna, R. Fender, T. Belloni, L. Stella, MNRAS **439**, L65 (2014)
232. A. Müller, M. Camenzind, A&A **413**, 861 (2004)
233. M.P. Muno, E.H. Morgan, R.A. Remillard, ApJ **527**, 321 (1999)
234. T. Nagao, R. Maiolino, A. Marconi, A&A **447**, 863 (2006)
235. K. Nandra, I.M. George, R.F. Mushotzky, T.J. Turner, T. Yaqoob, ApJ **477**, 602 (1997)
236. S. Nayakshin, C. Power, A.R. King, ApJ **753**, 15 (2012)
237. S. Nayakshin, D. Kazanas, T.R. Kallman, ApJ **537**, 833 (2000)
238. R. Narayan, I. Yi, ApJ **428**, L13 (1994)
239. R. Narayan, I. Yi, ApJ **444**, 231 (1995)
240. R. Narayan, J.E. McClintock, MNRAS **419**, L69 (2012)
241. J. Neilsen, J. Homan, ApJ **750**, 27 (2012)
242. J. Neilsen, J.C. Lee, Nature **458**, 481 (2009)
243. S.C. Noble, J.H. Krolik, J.F. Hawley, ApJ **711**, 959 (2010)
244. S.C. Noble, J.H. Krolik, J.F. Hawley, ApJ **692**, 411 (2009)
245. M.A. Nowak, B.A. Vaughan, MNRAS **280**, 227 (1996)
246. I.D. Novikov, K.S. Thorne, blho.conf, 343 (1973)
247. M.A. Nowak, B.A. Vaughan, J. Wilms, J.B. Dove, M.C. Begelman, ApJ **510**, 874 (1999)
248. J.A. Orosz, J.E. McClintock, J.P. Aufdenberg, R.A. Remillard, M.J. Reid, R. Narayan, L. Gou, ApJ **742**, 84 (2011)
249. J.A. Orosz, P.H. Hauschildt, A&A **364**, 265 (2000)
250. B. Paczyński, astro (2000). arXiv:astro-ph/0004129
251. B. Paczyńsky, P.J. Wiita, A&A **88**, 23 (1980)
252. D.N. Page, K.S. Thorne, ApJ **191**, 499 (1974)
253. J.C.B. Papaloizou, D.N.C. Lin, ApJ **438**, 841 (1995)
254. M.L. Parker et al., MNRAS **443**, 1723 (2014)
255. M.L. Parker et al., ApJ **808**, 9 (2015)
256. A.R. Patrick, J.N. Reeves, D. Porquet, A.G. Markowitz, V. Braito, A.P. Lobban, MNRAS **426**, 2522 (2012)
257. R.F. Penna, J.C. McKinney, R. Narayan, A. Tchekhovskoy, R. Shafee, J.E. McClintock, MNRAS **408**, 752 (2010)
258. R. Penrose, NCimR **1**, 252 (1969)
259. G. Ponti, R.P. Fender, M.C. Begelman, R.J.H. Dunn, J. Neilsen, M. Coriat, MNRAS **422**, L11 (2012)
260. J. Poutanen, G. Lipunova, S. Fabrika, A.G. Butkevich, P. Abolmasov, MNRAS **377**, 1187 (2007)
261. J.E. Pringle, MNRAS **258**, 811 (1992)
262. D. Proga, T.R. Kallman, ApJ **565**, 455 (2002)
263. D. Proga, T.R. Kallman, ApJ **616**, 688 (2004)
264. S. Puccetti, F. Fiore, G. Risaliti, M. Capalbi, M. Elvis, F. Nicastro, MNRAS **377**, 607 (2007)
265. J.C. Raymond, ApJ **412**, 267 (1993)
266. K.P. Rauch, R.D. Blandford, ApJ **421**, 46 (1994)
267. R.C. Reis, A.C. Fabian, R.R. Ross, J.M. Miller, MNRAS **395**, 1257 (2009)
268. R.C. Reis, M.T. Reynolds, J.M. Miller, D.J. Walton, D. Maitra, A. King, N. Degenaar, ApJ **778**, 155 (2013)
269. R.C. Reis, M.T. Reynolds, J.M. Miller, D.J. Walton, Nature **507**, 207 (2014)
270. R.A. Remillard, J.E. McClintock, ARA&A **44**, 49 (2006)
271. C.S. Reynolds, SSRv **183**, 277 (2014)
272. C.S. Reynolds, A.C. Fabian, ApJ **675**, 1048 (2008)
273. C.S. Reynolds, M.C. Begelman, ApJ **488**, 109 (1997)

274. C.S. Reynolds, A.J. Young, M.C. Begelman, A.C. Fabian, ApJ **514**, 164 (1999)
275. C.S. Reynolds, MNRAS **286**, 513 (1997)
276. C.S. Reynolds, L.W. Brenneman, A.M. Lohfink, M.L. Trippe, J.M. Miller, A.C. Fabian, M.A. Nowak, ApJ **755**, 88 (2012)
277. G. Risaliti, E. Nardini, M. Elvis, L. Brenneman, M. Salvati, MNRAS **417**, 178 (2011)
278. G. Risaliti et al., Nature **494**, 449 (2013)
279. G. Risaliti, M. Elvis, F. Nicastro, ApJ **571**, 234 (2002)
280. G. Risaliti, M. Elvis, S. Bianchi, G. Matt, MNRAS **406**, L20 (2010)
281. G. Risaliti, E. Nardini, M. Salvati, M. Elvis, G. Fabbiano, R. Maiolino, P. Pietrini, G. Torricelli-Ciamponi, MNRAS **410**, 1027 (2011)
282. G. Risaliti et al., ApJ **696**, 160 (2009)
283. R.R. Ross, A.C. Fabian, MNRAS **261**, 74 (1993)
284. R.R. Ross, A.C. Fabian, A.J. Young, MNRAS **306**, 461 (1999)
285. R.R. Ross, A.C. Fabian, MNRAS **358**, 211 (2005)
286. R.R. Ross, ApJ **233**, 334 (1979)
287. R.R. Ross, R. Weaver, R. McCray, ApJ **219**, 292 (1978)
288. R.R. Ross, A.C. Fabian, W.N. Brandt, MNRAS **278**, 1082 (1996)
289. A. RóżaNska, B. Czerny, AcA **46**, 233 (1996)
290. A. RóżaNska, MNRAS **308**, 751 (1999)
291. T.D. Russell et al., MNRAS **450**, 1745 (2015)
292. D.M. Russell, E. Gallo, R.P. Fender, MNRAS **431**, 405 (2013)
293. A. Sądowski, M. Abramowicz, M. Bursa, W. Kluźniak, J.-P. Lasota, A. Różańska, A&A **527**, A17 (2011)
294. A. Sadowski, M.A. Abramowicz, M. Bursa, W. Kluźniak, A. Różańska, O. Straub, A&A **502**, 7 (2009)
295. A. Sądowski, ApJS **183**, 171 (2009)
296. J.D. Schnittman, E. Bertschinger, ApJ **606**, 1098 (2004)
297. J.D. Schnittman, J.H. Krolik, ApJ **701**, 1175 (2009)
298. R. Shafee, R. Narayan, J.E. McClintock, ApJ **676**, 549 (2008)
299. R. Shafee, J.C. McKinney, R. Narayan, A. Tchekhovskoy, C.F. Gammie, J.E. McClintock, ApJ **687**, L25 (2008)
300. R. Shafee, J.E. McClintock, R. Narayan, S.W. Davis, L.-X. Li, R.A. Remillard, ApJ **636**, L113 (2006)
301. N.I. Shakura, R.A. Sunyaev, A&A **24**, 337 (1973)
302. N.I. Shakura, R.A. Sunyaev, MNRAS **175**, 613 (1976)
303. S.L. Shapiro, A.P. Lightman, ApJ **204**, 555 (1976)
304. T. Shimura, F. Takahara, ApJ **445**, 780 (1995)
305. V. Sochora, V. Karas, J. Svoboda, M. Dovčiak, MNRAS **418**, 276 (2011)
306. R. Speith, H. Riffert, H. Ruder, CoPhC **88**, 109 (1995)
307. J.F. Steiner, J.E. McClintock, R.A. Remillard, L. Gou, S. Yamada, R. Narayan, ApJ **718**, L117 (2010)
308. J.F. Steiner et al., MNRAS **416**, 941 (2011)
309. J.F. Steiner, J.E. McClintock, ApJ **745**, 136 (2012)
310. J.F. Steiner, J.E. McClintock, R. Narayan, ApJ **762**, 104 (2013)
311. J.F. Steiner, J.E. McClintock, J.A. Orosz, R.A. Remillard, C.D. Bailyn, M. Kolehmainen, O. Straub, ApJ **793**, L29 (2014)
312. L. Stella, Nature **344**, 747 (1990)
313. L. Stella, M. Vietri, ApJ **492**, L59 (1998)
314. L. Stella, M. Vietri, S.M. Morsink, ApJ **524**, L63 (1999)
315. L. Stella, M. Vietri, Phsy. Rev. Lett. **82**, 17 (1999)
316. O. Straub et al., A&A **533**, A67 (2011)
317. O. Straub, C. Done, M. Middleton, A&A **553**, A61 (2013)
318. L. Sun, X. Shu, T. Wang, ApJ **768**, 167 (2013)
319. J. Svoboda, M. Dovčiak, R. Goosmann, V. Karas, A&A **507**, 1 (2009)

320. M. Tagger, R. Pellat, A&A **349**, 1003 (1999)
321. Y. Tanaka et al., Nature **375**, 659 (1995)
322. C.B. Tarter, W.H. Tucker, E.E. Salpeter, ApJ **156**, 943 (1969)
323. A. Tchekhovskoy, J.C. McKinney, MNRAS **423**, L55 (2012)
324. Y. Terashima, N. Kamizasa, H. Awaki, A. Kubota, Y. Ueda, ApJ **752**, 154 (2012)
325. K.S. Thorne, ApJ **191**, 507 (1974)
326. Y. Ueda, K. Yamaoka, R. Remillard, ApJ **695**, 888 (2009)
327. P. Uttley, E.M. Cackett, A.C. Fabian, E. Kara, D.R. Wilkins, A&ARv **22**, 72 (2014)
328. P. Uttley, I.M. McHardy, S. Vaughan, MNRAS **359**, 345 (2005)
329. van der Klis, in *Proceedings of the NATO Advanced Study Institute Timing Neutron Stars.* NATO ASI Series C, Vol. 262. Kluwer, Dordrecht (1988)
330. H. van der Laan, Nature **211**, 1131 (1966)
331. B.A. Vaughan, M.A. Nowak, ApJ **474**, L43 (1997)
332. M. Volonteri, M. Sikora, J.-P. Lasota, A. Merloni, ApJ **775**, 94 (2013)
333. M. Volonteri, P. Madau, E. Quataert, M.J. Rees, ApJ **620**, 69 (2005)
334. D.J. Walton, E. Nardini, A.C. Fabian, L.C. Gallo, R.C. Reis, MNRAS **428**, 2901 (2013)
335. D.J. Walton, M.T. Reynolds, J.M. Miller, R.C. Reis, D. Stern, F.A. Harrison, ApJ **805**, 161 (2015)
336. C. Warner, F. Hamann, M. Dietrich, ApJ **608**, 136 (2004)
337. K.-Y. Watarai, J. Fukue, M. Takeuchi, S. Mineshige, PASJ **52**, 133 (2000)
338. R. Wijnands, J. Homan, M. van der Klis, ApJ **526**, L33 (1999)
339. D.R. Wilkins, A.C. Fabian, MNRAS **424**, 1284 (2012)
340. D.R. Wilkins, L.C. Gallo, MNRAS **449**, 129 (2015)
341. J.-H. Woo, C.M. Urry, ApJ **579**, 530 (2002)
342. A.J. Young, C.S. Reynolds, ApJ **529**, 101 (2000)
343. A.J. Young, R.R. Ross, A.C. Fabian, MNRAS **300**, L11 (1998)
344. S.N. Zhang, W. Cui, W. Chen, ApJ **482**, L155 (1997)
345. Y. Zhu, S.W. Davis, R. Narayan, A.K. Kulkarni, R.F. Penna, J.E. McClintock, MNRAS **424**, 2504 (2012)
346. A. Zoghbi, A.C. Fabian, P. Uttley, G. Miniutti, L.C. Gallo, C.S. Reynolds, J.M. Miller, G. Ponti, MNRAS **401**, 2419 (2010)
347. A. Zoghbi, A.C. Fabian, C.S. Reynolds, E.M. Cackett, MNRAS **422**, 129 (2012)
348. A. Zoghbi, C. Reynolds, E.M. Cackett, G. Miniutti, E. Kara, A.C. Fabian, ApJ **767**, 121 (2013)
349. A. Zoghbi et al., ApJ **789**, 56 (2014)
350. P.T. Zycki, J.H. Krolik, A.A. Zdziarski, T.R. Kallman, Apj **437**, 597 (1994)

Chapter 4
Winds from Black Hole Accretion Flows: Formation and Their Interaction with ISM

Feng Yuan

Abstract Black hole hot accretion flows occur in the regime of relatively low accretion rates and are operating in the nuclei of most of the galaxies in the universe. In this chapter, I will review one of the most important progresses in recent years in this field, which is about the wind or outflow. This progress is mainly attributed to the rapid development of numerical simulations of accretion flows, combined with observations on, e.g., Sgr A*, the supermassive black hole in the Galactic center. The following topics will be covered: theoretically why do we believe strong winds exist; where and how are they produced and accelerated; what are their main properties such as mass flux and terminal velocity; the comparison of the properties between wind and "disk-jet"; the main observational evidences for wind in Sgr A*; and one observational manifestation of the interaction between wind and interstellar medium, namely the formation of the Fermi bubbles in the Galactic center.

4.1 Introduction

Black hole accretion is a fundamental physical process in the universe. It is the standard model for the central engine of active galactic nuclei (AGNs), and also plays a central role in the study of black hole X-ray binaries, Gamma-ray bursts, and tidal disruption events. According to the temperature of the accretion flow, the accretion models can be divided into two classes, namely cold and hot. The standard thin disk model belongs to the cold disk, since the temperature of the gas is far below the virial value Shakura and Sunyaev [45] (see reviews by [1, 6, 20, 40]). The disk is geometrically thin but optically thick and radiates multi-temperature black body spectrum. The radiative efficiency is high, ~ 0.1, independent of the accretion rate. The model has been successfully applied to luminous sources such as luminous AGNs and black hole X-ray binaries in the thermal state.

F. Yuan (✉)
Shanghai Astronomical Observatory, Chinese Academy of Sciences,
80 Nandan Road, Shanghai 200030, China
e-mail: fyuan@shao.ac.cn

© Springer-Verlag Berlin Heidelberg 2016
C. Bambi (ed.), *Astrophysics of Black Holes*, Astrophysics
and Space Science Library 440, DOI 10.1007/978-3-662-52859-4_4

153

In addition to the cold disk model, the accretion equations have another set of solutions, i.e., the hot accretion flow. The most well-known and pioneer hot accretion flow model is the advection-dominated accretion flow [2, 37, 38]. This model applies when the mass accretion rate is below $\sim(0.1 - 0.3)\alpha^2\dot{M}_{Edd}$, with $\dot{M}_{Edd} \equiv 10L_{Edd}/c^2$ is defined as the Eddington accretion rate and α is the viscous parameter. Above this critical rate but below $\sim\alpha\dot{M}_{Edd}$, we have another hot accretion flow model called luminous hot accretion flows [54]. In contrast to the cold disk, the temperature of the hot accretion flow is much higher, close to the virial value. The flow is thus geometrically thick. The radiative efficiency of hot accretion flow is very low when the accretion rate is low, but quickly increase with the increasing accretion rates [53]. Hot accretion flow is operating in the nuclei of perhaps most galaxies in the universe, and the quiescent and hard states of black hole X-ray binaries. Yuan and Narayan [58] present a comprehensive review of various aspects of the model, including the one-dimensional and multi-dimensional dynamics, radiation, jet formation, wind, and the applications in Sgr A*, low-luminosity AGNs, black hole X-ray binaries, and AGN feedback.

For a long time, the mass accretion rate of accretion flow, both cold and hot, is assumed to be a constant of radius. This implies that all the gas available at the outer boundary of the accretion flow will be accreted into the black hole horizon, only except at the innermost region very close to the black hole where a jet is formed. However, in recent years, we have realized the existence of strong wind from black holt accretion flow in both observational and theoretical aspects. We now understand that most of the gas available at the outer boundary will be lost in wind rather than accreted into the black hole horizon. In fact, the study of wind has become a hot topic in the field of black hole accretion. The reason is twofold. The first is that this is crucial for our understanding of the dynamics of accretion. The second reason is that this can help us to understand observations on wind. The third reason is related to AGN feedback. AGN feedback is now believed to play an important role in galaxy formation and evolution [18, 27, 28]. There are two kinds of medium for the interaction between an AGN and the interstellar medium, i.e., photons and matter. The latter includes wind and jet. So to understand AGN feedback, we need to first understand the main properties of wind.

In this chapter, I will review our current understanding of the various aspects of wind from hot accretion flows.

4.2   Formation of Wind from a Hot Accretion Flow

The study of wind from hot accretion flows started from about 15 years ago. This is young compared with the history of the study of wind from a thin disk, which started more than twenty years ago (e.g., [41]). However, since in many cases we do not need to take into account radiation, and the hot accretion flow is technically much easier to simulate than a thin disk, we have a better understanding to the wind launched from hot accretion flows.

4.2.1 Brief History of Study of Wind from Hot Accretion Flows

One of the most important pioneer works in this aspect is Stone, Pringle and Begelman [7] (see also [23, 24]). They performed the first global hydrodynamical numerical simulation of hot accretion flow and calculated the following time-averaged radial profiles of inflow and outflow rates,

$$\dot{M}_{\text{in}}(r) = 2\pi r^2 \left\langle \int_0^{\pi} \rho \min(v_r, 0) \sin\theta d\theta \right\rangle_{t\phi}, \tag{4.1}$$

$$\dot{M}_{\text{out}}(r) = 2\pi r^2 \left\langle \int_0^{\pi} \rho \max(v_r, 0) \sin\theta d\theta \right\rangle_{t\phi}, \tag{4.2}$$

where the angle brackets represent time averages (and also average over the azimuthal angle ϕ in the case of 3D simulations). Note that the order of doing time-average and the integral will make significant differences. Note also that the outflow rate calculated by Eq. (4.2) does not necessarily represent the mass flux of "real outflow", because the positive radial velocity may just come from the turbulent motion of the accretion flow. The most important result they obtained is that the inflow rate based on Eq. (4.1) follows a power-law function of radius,

$$\dot{M}_{\text{in}}(r) = \dot{M}_{\text{in}}(r_{\text{out}}) \left(\frac{r}{r_{\text{out}}} \right)^s. \tag{4.3}$$

Here $\dot{M}_{\text{in}}(r_{\text{out}})$ is the mass inflow rate at the outer boundary r_{out}. The dynamical range of this simulation is not large, spanning less than two orders of magnitude in radius. The results were later confirmed by simulations with a much larger radial dynamical range of four orders of magnitude [56]. Moreover, MHD simulations yield very similar results (e.g., [26, 39, 46]). In almost all cases, we typically have $s \sim 0.5 - 1$ (see review in [56]).

The predicted inward decrease of accretion rate has soon been confirmed by two observations, both are on Sgr A*. One is the detection of radio polarization at a level of $2 - 9\%$ (e.g., [5, 12, 31]). Such high polarization requires that the mass accretion rate close to the black hole horizon must be within a certain range, which is two orders of magnitude lower than the Bondi rate obtained from *Chandra* observations [43]. The other observational evidence is from the *Chandra* observation of the iron emission lines originated from the hot accretion flow [52]. The modeling to the $K\alpha$ lines indicates a flat radial density profile around the Bondi radius, confirming that the mass accretion rate decreases with decreasing radius. This is because, if the mass accretion rate were a constant of radius, the density profile would be much steeper (Fig. 4.1).

Now the question is, what is the reason of the inward decrease of mass accretion rate? Two competing models have been proposed. In the adiabatic inflow–outflow

Fig. 4.1 Radial profiles of mass inflow rate $\dot{M}_{in}$, mass outflow rate $\dot{M}_{out}$ (Eqs. 4.1 and 4.2), and net mass accretion rate $\dot{M}_{net}$ defined as their difference. *Top* Results from a two-dimensional Newtonian HD simulation of a hot accretion flow (Taken from Stone et al. [47]). *Solid, dashed,* and *dotted lines* correspond to $\dot{M}_{in}$, $\dot{M}_{out}$, and $\dot{M}_{net}$, respectively. *Bottom Solid lines* indicate equivalent results from a three-dimensional GRMHD simulation of a hot accretion flow around a non spinning *black* hole (Taken from Yuan and Narayan [58]). *Dashed lines* indicate results for a different kind of time averaging, as described in the text. The true mass outflow rate is ∼60 % of the *solid green lines* [59]

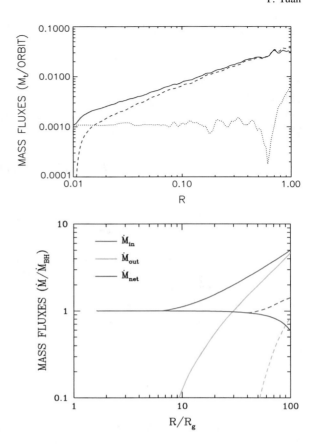

solution (ADIOS), the inward decrease of mass accretion rate is due to the mass lost in the wind [7, 8, 11]. The other model is the convection-dominated accretion flow (CDAF) model. In this model, the accretion flow is assumed to be convectively unstable.[1] The inward decrease of accretion rate is explained as more and more gas is locked in convective eddies during accretion [3, 25, 35, 42]. For a long time, it is unclear which scenario is physical.

To investigate this problem, numerical simulations have been performed [29, 36, 56]. Both Yuan, Bu and Wu [56] and Narayan et al. [36] found that in the presence of magnetic field, the hot accretion flow is convectively *stable*. This indicates that the CDAF model may not apply, leaving outflow/wind as the only possible solution. The fundamental question is, how strong the wind is. Narayan et al. [36] calculated the outflow rate based on Eq. (4.2), except that they move the $t\phi$ average inside the integral. The advantage of this approach is that it eliminates contributions from

---

[1] As we will introduce in the next paragraph, this assumption is likely true only when the magnetic field is not included in the analysis of the stability.

turbulent motion. The disadvantage is that, as shown in Yuan et al. [59] and also described in the present paper later, it also eliminates significant mass flux of real outflow. Consequently, they found substantially lower outflow rate than Eq. (4.2). In fact, only upper limit was reported in Narayan et al. [36] since the outflow rate was found to not converge with time.

On the other hand, [56] obtained a different result and showed that the mass flux of outflow should be large, i.e., being a significant fraction of that described by Eq. (4.2). This conclusion is mainly based on the following argument. That is, if the main contributor of Eq. (4.2) is turbulence, we would expect that the properties of inflow and outflow, such as angular momentum and temperature, should be roughly same; while they found that they are quite different. For an example, the specific angular momentum of outflow is much higher than that of the inflow. The hydrodynamical simulations by [29] also found strong outflow as Yuan, Bu and Wu [56].

To obtain the mass flux of wind, the crucial point is how to discriminate the real wind and turbulent motion. Yuan et al. [59] finally solved this issue by using a "trajectory" approach, combined with the data of three dimensional GRMHD numerical simulation of accretion. Different from the streamline analysis often adopted in accretion literature, this approach can provide the trajectory of each "virtual test particle" in the accretion flow and thus directly show whether the flow is turbulent outflow or real outflow. Using this approach, they found that the mass flux of wind is quite strong. In fact, the mass flux of winds is $\sim 60\%$ of that calculated by Eq. (4.2) (see Sect. 2.2 for details). The rather weak outflow rate obtained in Narayan et al. [36] is because outflow is intrinsically instantaneous. The outflow stream can wander around in 3D space thus will be cancelled if the time-average is performed first. Their work indicates that the mass lost via the wind is the reason for the inward decrease of the accretion rate (Eq. 4.3). In the following we summarize the main results they have obtained.

4.2.2 Main Properties of Winds

The trajectories of some virtual test particles obtained in Yuan et al. [59] are shown in Fig. 4.2. We can see that in the main body of the accretion flow around the equatorial plane, it is inflow, and the motion is quite turbulent. Winds are evident in the coronal region and their motion is much less turbulent. Below we summarize their main properties, including the mass flux, poloidal speed, fluxes of energy and momentum.

Mass Flux

Figure 4.3 shows various mass flow rates as a function of radius. Black and blue solid lines show the inflow and outflow rates calculated following Eqs. (4.1 and 4.2) while the blue solid line shows their difference. The red and blue dashed lines show the mass fluxes of wind and real inflow (i.e., excluding the turbulent inflow) at a certain time of the simulation. The mass flux of wind can be described by

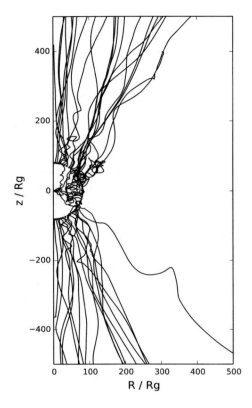

Fig. 4.2 Lagrangian trajectories of the "test particles" originating from r = 80r_g (the *black circle* in the figure) in the 2D ($r - \theta$) plane. Winds are evident in the coronal region. The inflow concentrates within the main disk body around the equatorial plane, and their motion is turbulent. Taken from Yuan et al. [59]

$$\dot{M}_{\mathrm{wind}}(r) = \dot{M}_{\mathrm{BH}} \left(\frac{r}{20r_s} \right)^s , \quad s \approx 1, \tag{4.4}$$

where $\dot{M}_{\mathrm{BH}}$ is the mass accretion rate at the black hole horizon and $r_s \equiv 2GM/c^2$ is the Schwarzschild radius. Comparing it with the outflow rate $\dot{M}_{\mathrm{out}}$ calculated by Eq. (4.2), we find that the mass flux of wind is $\dot{M}_{\mathrm{wind}} \sim 60 \% \dot{M}_{\mathrm{out}}$. This confirms the conclusion obtained by Yuan, Bu and Wu [56]. Also shown in the figure are the mass fluxes of the wind calculated following the Narayan et al. [36] method, which is much weaker, equal to $\dot{M}_{\mathrm{BH}}$ only until 50r_s.

How large can the value of r in Eq. (4.4) be? This is important since this will determine the total wind flux. [14, 15] studied this problem by using two dimensional hydrodynamical and MHD numerical simulations. They focus on the large radius of the accretion flow. At that region, in addition to the central black hole, the nuclear star cluster also contributes to the gravitational potential thus needs to be taken into account. The velocity dispersion of stars is assumed to be a constant and the gravitational potential of the nuclear star cluster $\phi \propto \sigma^2 \ln(r)$, where σ is the velocity dispersion of stars and r is the distance from the center of the galaxy. It is found that when the gravity is dominated by the nuclear star cluster, i.e., $r > r_A \equiv GM/\sigma^2$,

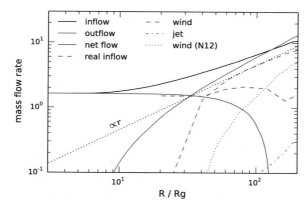

Fig. 4.3 Various mass flow rates as a function of radius. *Black* and *red solid lines* show the total inflow and outflow rates calculated following Eqs. (4.1–4.2), while the *blue solid line* denotes their difference. The *red* and *blue dashed lines* denote the mass flux of the wind and real inflow calculated at a certain time. The *red dot-dashed line* shows the mass flux of the disk jet. For comparison, the *red dotted line* shows the mass flux of wind calculated following the method in Narayan et al. [36]. Taken from Yuan et al. [59]

wind launched from the accretion flow almost disappear. This result indicates that there exists an upper limit of the value of r in Eq. (4.4), which is $r < r_A$. Since σ is close to the sound speed of the accreting gas around the Bondi radius, the value of r_A is close to the Bondi radius, which is the outer boundary of the accretion flow.

Poloidal Velocity of Wind and Disk Jet

We are mainly interested in the poloidal speed of wind. This is because, at large radius, the poloidal speed of wind is the dominant component since the magnetic field becomes subdominant. Figure 4.4 shows the angular distribution of the poloidal speed of wind at radius $r = 160r_g$. The results at the other radius are similar. We can see that the poloidal speed as a function of θ has a sharp jump at $\theta \sim 15°$ away from the rotation axis. The poloidal speed of outflow close to the axis is $>0.3c$, much larger than that away from the axis, which is $<0.05c$. The outflow within $\theta < 15°$ to the axis is thus naturally identified as the "disk jet", while the outflow out of this range is identified as wind. Note that the simulation data are for a nonrotating black hole, so the presence of the disk jet is irrelevant to black hole spin, although the spin of the black hole may strengthen the disk jet. The disk jet originates from the inner region of the disk and is powered by the rotation energy of the accretion flow (see more discussion below). This is different from the Blandford–Znajek jet originated from the black hole horizon, which is powered by the spin energy of the black hole [10]. Other main differences between the two types of jet are that the disk jet is sub-relativistic and matter dominated, while the Blandford–Znajek jet is relativistic and Poynting flux dominated (see [58] for a summary).

Fig. 4.4 Poloidal speed (in units of speed of light) of wind as a function of θ at $r = 160r_g$ and various ϕ. The values of ϕ are denoted by the color of the lines. We can see that close to the axis, the poloidal speed is much larger than in other regions. We identify this part as the disk jet. Taken from Yuan et al. [59]

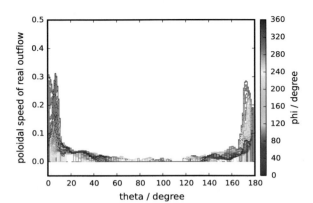

The mass flux-weighted poloidal speed of wind at radius r is described by

$$v_{\mathrm{p,wind}} \approx 0.2 v_k(r) \tag{4.5}$$

On the one hand, the wind can be launched from any radius r. On the other hand, we see in Fig. 4.3 that the wind mass flux increases rapidly with radius. Therefore, this result primarily reflects wind launched close to radius r.

An important question is the evolution of the poloidal speed of wind along their trajectories. For disk jet, their poloidal speed significantly increases; while for wind, the poloidal speed roughly keeps constant or slightly increases along the trajectory. If there were no acceleration during the propagation of wind, the poloidal speed would decrease outward because of the gravity. Therefore, this result implies that the winds are accelerated along the trajectory. The acceleration forces will be discussed in Sect. 2.3. In summary, the terminal speed of wind originated from radius r can be described by Yuan et al. [59] (see also [57])

$$v_{\mathrm{p,term}} \approx (0.2 - 0.4) v_k(r) \tag{4.6}$$

The above behavior of poloidal velocity of wind implies that the Bernoulli parameter of wind Be is not constant but increases. Note that only for strictly steady and inviscid flow Be is a constant along the trajectory, while a real accretion flow is always turbulent.

Fluxes of Energy and Momentum

The fluxes of energy and momentum of wind and disk jet are calculated as follows,

$$\dot{E}_{\mathrm{jet(wind)}}(r) = \int \frac{1}{2}\rho(r,\theta,\phi)v_p^3(r,\theta,\phi)r^2 \sin(\theta)d\theta d\phi, \tag{4.7}$$

Fig. 4.5 The radial profile of the energy fluxes of wind (*red solid*) and disk jet (*blue dashed*) for a non-rotating black hole. Taken from [59]

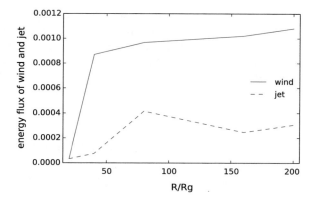

$$\dot{P}_{\text{jet(wind)}}(r) = \int \rho(r, \theta, \phi) v_p^2(r, \theta, \phi) r^2 \sin(\theta) d\theta d\phi. \tag{4.8}$$

The integration over θ for wind and the disk-jet is bounded by $\theta \approx 15°$ according to discussions above. Figure 4.5 shows the energy flux of wind and disk jet as a function of radius. We see that the energy flux of the wind is >3 times stronger than that of the disk jet, while the contrast in the momentum flux between wind and jet is even larger [59]. This result is mainly because of the low density in the disk jet. Obviously, it implies the importance of wind in comparison with disk jet in the context of AGN feedback. Of course, this is for a non-rotating black hole. In the case of a spinning black hole, the relative importance of a jet in comparison with winds will become stronger ([44][2]; Yuan et al. in preparation). For $r > 40r_g$, we have

$$\dot{E}_{\text{wind}}(r) \approx \frac{1}{2}\dot{M}_{\text{wind}}(r)v_{p,\text{wind}}^2(r) \approx \frac{1}{1000}\dot{M}_{\text{BH}}c^2. \tag{4.9}$$

This result indicates that the energy flux at large radius is roughly saturated, consistent with Fig. 4.5. The main reason energy flux saturates is $s = 1$ in Eq. (4.4).

The energy flux obtained in Eq. (4.9) is in good agreement with that required in large scale AGN feedback simulations (e.g., [16]). In these works, AGN feedback is involved to heat the inter-cluster medium to compensate for rapid cooling rate in the systems (i.e., the cooling flow problem). It was found that to be consistent with observations of both isolated galaxies and galaxy clusters, the required "mechanical feedback efficiency", defined as $\varepsilon \equiv \dot{E}_{\text{wind}}/\dot{M}_{\text{BH}}c^2$, must be in the range of $\sim 10^{-3} - 10^{-4}$. The result shown in Eq. (4.9) provides a natural explanation for this required value of ε.

In accretion systems with extremely low mass accretion rates such as Sgr A*, the density of the accretion flow is very low thus the mean free path of particles may

---

[2]In the calculations of many wind properties presented in Sadowski et al. [44] such as the mass flux of wind, they do the time-average first to the velocity field. Since wind is instantaneous, their result should be regarded as the lower limit.

be large compared with the typical length-scale of the system. In this case, thermal conduction will play an important role. Bu, Wu and Yuan [13] have studied the effect of thermal conduction on the properties of winds. They find that the mass flux of wind slightly increases due to the presence of thermal conduction, while the energy flux of wind increases by a factor of ~10 mainly because of the significant increase of wind velocity.

4.2.3 Acceleration Mechanism of Wind and Disk Jet

What are the mechanisms to accelerate the wind against the gravity? This can be studied by analyzing the forces in the wind and disk jet region [32, 59]. Figure 4.6 shows the results for three representative points in the disk jet, wind and main disk regions. For the wind, the main driving forces are the centrifugal force and the gradient of magnetic and gas pressure. From the figure we notice that the gradient of the magnetic pressure is "downward", pointing toward the positive θ direction. This surprising result reflects the strong fluctuation of the accretion flow. If we choose another time or another location to do the force analysis, we very likely find that the gradient of the magnetic pressure becomes "upward". The direction of the gas pressure gradient also strongly fluctuates with time and location. Therefore the acceleration is not a continuous process but stochastic [32, 59]. But statistically, the gradients of both the gas and magnetic pressure are pointing along the positive r direction thus are helpful to the acceleration of wind. Their magnitudes are also comparable to the centrifugal force, as shown by Fig. 4.6. The centrifugal force is important because the specific angular momentum of wind is large. For example, Yuan, Bu and Wu [56] find that the specific angular momentum of outflow is significantly larger than that of inflow. In that work, "outflow" includes both wind and turbulent outflow. Figure 4.7 shows the radial profile of the angular momentum of wind. We can see that the angular momentum of wind is actually super-Keplerian. Such a result is very likely because of the angular momentum transport by the magnetic field, i.e., the wind gas is somehow forced to co-rotate with the magnetic field line rooted in the main body of accretion flow, as described in the model of Blandford and Payne [9]. In summary, the mechanism of wind acceleration is similar to the Blandford and Payne [9] model in the sense that the magneto-centrifugal force play an important role. But different from the Blandford and Payne [9] mechanism, here we don't have a large scale ordered poloidal field and the other forces such as the gradient of gas pressure also play important roles.

For the disk jet, the dominant driving force is the gradient of the magnetic pressure. This is consistent with the "magnetic tower" model proposed by Lynden–Bell [30].

Fig. 4.6 Force analysis at three representative locations corresponding to the disk jet, wind and the main body of the accretion disk. The *arrows* indicate force direction, while length represents force magnitude. Taken from Yuan et al. [59]. See also Moller and Sadowski [32]

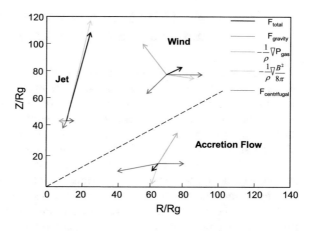

Fig. 4.7 The radial profiles of flux-weighted angular momentum (in unit of Keplerian angular momentum) of wind (*solid blue*), turbulent outflow (*solid black*), real inflow (*dashed blue*), and turbulent inflow (*dashed blue*). The angular momentum of wind is super-Keplerian

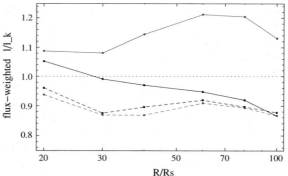

4.2.4   Why Do Winds Exist?

So far our discussion focuses only on whether strong wind exists in hot accretion flows and what are their properties. But why does strong wind exist or why some inflowing gas turn around and become outflow? In addition to above-mentioned works, this question is also addressed in Begelman [11] and [21]. For example, [21] argue that the energy equilibrium never can be reached in hot accretion flows unless wind is present and take away some energy. Based on Sect. 2.3, it looks that the wind is produced because of both "energy" and "momentum" reasons. The gradient of gas pressure corresponds to "energy driven". Angular momentum is transported from one fluid element to another perhaps by magnetic field, which makes the centrifugal force large thus the originally inflowing fluid is inclined to become into wind. The magneto-centrifugal force corresponds to "momentum-driven".

4.3   Interaction of Winds with Interstellar Medium: The Formation of the Fermi Bubbles

Using the Fermi-LAT, two giant gamma-ray Bubbles has been discovered, located above and below the Galactic plane [48]. The bubbles extend to $\sim50°$ above and below the Galactic plane, and the width is $\sim40°$ in longitude. Several models have been proposed to explain the formation of the Fermi bubbles. They often invoke the interaction between a jet (e.g., [22]) or wind and the interstellar medium (see [34] for a brief review). In different models, while the jet comes from the accretion flow, the wind can either come from the cold or hot accretion flows around the black hole [33, 34, 60, 61], or star formation [17]. In both the jet and accretion wind models, they have to assume that the luminosity of Sgr A* was several orders of magnitude higher in the last. There seems to be many observational evidences for it (see review in [50]). In the jet model, they must assume that the jet is perpendicular to the Galactic plane in order to explain the morphology of the bubbles which are perpendicular to the Galactic plane. As pointed out by Zubovas et al. [61], this is a strong assumption since statistical studies to other galaxies indicate that this is usually not the case. In addition, the mass lost rate in the jet is also assumed to be very large, close to or even larger than Eddington.

Mou et al. [33, 34] propose that the Fermi bubbles are produced by the interaction between the wind launched from the hot accretion flow around Sgr A* and the interstellar medium. The mass accretion rate of the accretion flow is determined by other independent observations [50], which is $\sim10^{-2}\dot{M}_{Edd}$, thus is not a free parameter. This rate is still well in the regime of the hot accretion flow [58] thus strong winds are expected, as described in the previous sections. The properties of wind such as its mass flux, terminal velocity, and angular distribution are all determined by the small-scale MHD numerical studies [59] thus are again not free parameters. In addition to thermal particles, Mou et al. [34] assume that relativistic electrons and protons also exist in the wind. These particles are likely produced by, e.g., magnetic reconnection process occurred in the coronal region of the accretion flow (e.g., [55]). These relativistic protons (also called cosmic ray protons; CRp) will collide with the thermal protons in the ISM and produce neutral pions; these poins will decay and produce gamma-ray photons of the Fermi bubbles. So in order to obtain the Gamma-ray radiation and compare it with observations, the main task is to calculate the spatial distribution of CRp. This is achieved in Mou et al. [34] by treating CRp as another kind of fluid and solving the time-dependent two-dimensional two-fluid equations using the simulation code *ZEUS*.

Figure 4.8 shows the simulated morphology of the bubble. The left and right plots show the number density and temperature of the thermal gas, respectively. The purple line in the figure denotes the contact discontinuity (CD) between the wind and the ISM. Near the CD, the CRp pressure is comparable to the thermal pressure of the shocked ISM, so it expels some thermal gas away from the CD, leaving a zone with density somewhat lower than the "typical" density of the shocked ISM. We call it a "permeated zone". The CRp and thermal protons are well mixed in this zone thus

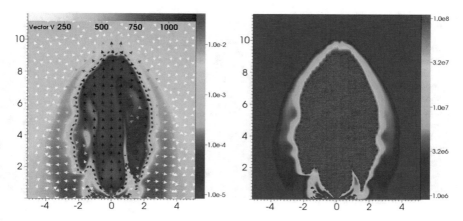

Fig. 4.8 $X-Z$ sectional views of the simulation results. Coordinates are in units of kpc. *Left* number density distribution of thermal electrons (n_e) in units of cm^{-3}. Velocity vectors are also plotted, with the color bar at the top of the plot denoting the value of velocity in units of km s^{-1}. *Right* temperature in units of Kelvin. The *dashed lines* in these three maps denotes the contact discontinuity (CD). Taken from [34]

$p-p$ collision is frequent. Therefore, most of the gamma-ray radiation is produced from this zone. We can see that the shape of this zone is roughly consistent with the observed shape of the Bubbles. The central molecular zone located in the galactic center plays an important role in collimating the wind and the formation of the bubble morphology. One caveat exists in their model. That is, although a jet should also exist for accretion rate of $10^{-2}\dot{M}_{\mathrm{Edd}}$, they assume that the interaction of the jet and ISM can be neglected since the jet may just pierce through the ISM in a narrow channel without depositing much energy in the ISM. This assumption seems to be reasonable as shown by the numerical simulation of Vernaleo and Reynolds [51].

The magnetic field in the shocked ISM is roughly parallel to the CD. However, the alignment is not perfect, which allows some CRp diffuse into the shocked ISM region and form the "permeated zone". But under such a kind of magnetic field configuration, CRs cannot diffuse too far away from the CD. Therefore the morphology of the bubbles is determined by the CD, and this is also the reason why the edges of the Fermi bubbles look sharp.

The processes considered in Mou et al. [34] for the production of gamma-ray emission include: (1) the production of neutral pions by the collision between thermal protons and CRp, which will further decay and produce gamma-ray; (2) the $p-p$ collision will also produce some charged pions which will generate second-order leptons. These leptons will scatter with the seed photons and produce gamma rays; (3) In the processes of CRp production such as magnetic reconnection in the disk corona, some CRe will also be produced. These electrons can also produce gamma-ray photons by scattering with seed photons.

Figure 4.9 shows the calculated spectral energy distribution, together with the observational data. The model can fit the data well. In addition to the spectrum, the

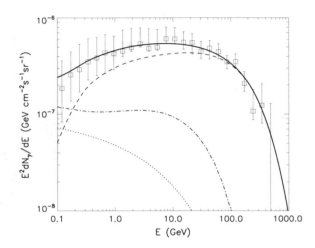

Fig. 4.9 The gamma-ray spectral energy distribution calculated based on run A. The rectangles with error bars show the latest observational results of Ackermann et al. [4]. The *solid line* is the sum of the *dashed* (for the π^0 decays), *dotted* (for IC process of the secondary leptons generated in hadronic reaction), and *dot-dashed* (for IC of the primary electrons) lines. Taken from Mou et al. [34]

model can also successfully explain the other main observational features of the Fermi bubbles: (1) limb-brightened surface brightness; (2) the width of the boundary; (3) the temperature of the shocked ISM just outside of the CD [49]; (4) the bulk motion velocity of the gas at the edge of the Fermi bubbles [19].

4.4 Summary

Numerical simulations have shown that the mass accretion rate decreases inward toward the black hole (Fig. 4.1). This theoretical result has been confirmed by observations. To understand this result, two models have been proposed, namely convection and outflow (or wind). Recent theoretical works have shown that it is the mass lost in the wind that results in the inward decrease of accretion rate (Fig. 4.2). Based on the MHD simulation data, we can follow the trajectories of the virtual particles thus precisely measure the various properties of wind, such as the mass flux (Fig. 4.3 and Eq. 4.4), poloidal speed (Fig. 4.4 and Eqs. 4.5 and 4.6.), the fluxes of energy and momentum (Fig. 4.5 and Eqs. 4.7 and 4.8). It is found that The Bernoulli parameter is usually not a constant and the poloidal speed of wind roughly keeps constant or increases along their trajectories. The acceleration forces are the centrifugal force, and the gradient of the gas and magnetic pressure (Fig. 4.6). It is therefore a combination of the magneto-centrifugal mechanism similar to that proposed by Blandford and Payne [9] and the thermal mechanism.

One interesting result is that the angular distribution of the poloidal speed of wind has a sharp jump at $\sim\theta = 15°$. Close to the axis, the speed is as high as $0.3c$ while beyond this region the speed is about ten time lower. The former is therefore identified as the "disk jet". The comparison between this kind of jet and the Blandford and Znajek jet and more importantly, which kmind of jet is physically associated with the observed "real" jet is interesting topics to be studied in the future.

The wind from hot accretion flow is difficult to be detected by the usual absorption line approach, because the temperature of wind gas is too high thus fully ionized. But the interaction of such wind with the interstellar medium may have some observational consequence and this process may result in the formation of the Fermi bubbles observed in the Galactic center by the *Fermi* telescope. This idea has been worked out in Mou et al. [33, 34] and the model has successfully explained the key observations such as the morphology and Gamma ray spectrum. The advantage of this model compared with the other models of the Fermi bubbles is that the key parameters of the model, i.e., the mass accretion rate in the accretion flow and the properties of wind are not free parameters. The former is adopted from the other independent observational constrains, while the properties of wind come from the MHD simulation of hot accretion flow as described in this review.

Acknowledgments This project was supported in part by the National Basic Research Program of China (973 Program, grant 2014CB845800), the Strategic Priority Research Program The Emergence of Cosmological Structures of CAS (grant XDB09000000), the Natural Science Foundation of China (grants 11133005 and 11573051), and the CAS/SAFEA International Partnership Program for Creative Research Teams.

References

1. M.A. Abramowicz, P.C. Fragile, Living Reviews in Relativity **16**, 1 (2013)
2. M.A. Abramowicz, X. Chen, S. Kato, J.P. Lasota, O. Regev, ApJ **438**, L37 (1995)
3. M.A. Abramowicz, I.V. Igumenshchev, E. Quataert, R. Narayan, ApJ **565**, 1101 (2002)
4. M. Ackermann, A. Albert, W.B. Atwood, et al., ApJ **793**, 64 (2014)
5. D.K. Aitken, J. Greaves, A. Chrysostomou et al., ApJ **534**, L173 (2000)
6. O. Blaes, Spac. Sci. Rev. **183**, 21 (2014)
7. R.D. Blandford, M.C. Begelman, MNRAS **303**, L1 (1999)
8. R.D. Blandford, M.C. Begelman, MNRAS **349**, 66 (2004)
9. R.D. Blandford, D.G. Payne, MNRAS **199**, 883 (1982)
10. R.D. Blandford, R.L. Znajek, MNRAS **179**, 433 (1977)
11. M.C. Begelman, MNRAS **420**, 2912 (2012)
12. G.C. Bower, M.C.H. Wright, H. Falcke, D.C. Backer, ApJ **588**, 331 (2003)
13. D.F. Bu, M.C. Wu, Y.F. Yuan, MNRAS **459**, 746 (2016)
14. D.F. Bu, F. Yuan, Z. Gan, X.H. Yang, ApJ **813**, 83 (2016a)
15. D.F. Bu, F. Yuan, Z. Gan, X.H. Yang, ApJ **823**, 90 (2016b)
16. L. Ciotti, J.P. Ostriker, D. Proga, ApJ **717**, 708 (2010)
17. R.M. Crocker, F. Aharonian, PhRvL **106**, 101102 (2011)
18. A.C. Fabian, ARA&A **50**, 455 (2012)
19. T. Fang, X. Jiang, ApJL **785**, L24 (2014)
20. J. Frank, A. King, D.J. Raine, *Accretion Power in Astrophysics* (Cambridge University Press, Cambridge, 2002)
21. W.M. Gu, ApJ **799**, 71 (2015)
22. F. Guo, W.G. Mathews, ApJ **756**, 181 (2012)
23. I.V. Igumenshchev, M.A. Aramowicz, ApJ **537**, L27 (1999)
24. I.V. Igumenshchev, M.A. Aramowicz, ApJS **130**, 463 (2000)

25. I.V. Igumenshchev, ApJ **577**, L31 (2002)
26. I.V. Igumenshchev, R. Narayan, M.A. Abramowicz, ApJ **592**, 1042 (2003)
27. A. King, K. Pounds, ARA&A **53**, 115 (2015)
28. J. Kormendy, L.C. Ho, ARA&A **51**, 511 (2013)
29. J. Li, J. Ostriker, R. Sunyaev, ApJ **767**, 105 (2013)
30. D. Lynden-Bell, MNRAS **341**, 1360 (2003)
31. D.P. Marrone, J.M. Moran, J.H. Zhao, R. Rao, ApJ **654**, 57 (2007)
32. A. Moller, A. Sadowski (2015) ApJ submitted (arXiv:1509.06644)
33. G. Mou, F. Yuan, D. Bu et al., ApJ **790**, 109 (2014)
34. G. Mou, F. Yuan, Z. Gan, M. Sun, ApJ **811**, 37 (2015)
35. R. Narayan, I.V. Igumenshchev, M.A. Abramowicz, ApJ **539**, 798 (2000)
36. R. Narayan, A. Sädowski, R.F. Penna, A.K. Kulkarni, MNRAS **426**, 3241 (2012)
37. R. Narayan, I. Yi, ApJ **428**, L13 (1994)
38. R. Narayan, I. Yi, ApJ **452**, 710 (1995)
39. U.L. Pen, C.D. Matzner, S. Wong, ApJ **596**, L207 (2003)
40. J.E. Pringle, ARA&A **19**, 137 (1981)
41. D. Proga, ASP Conference Series, in *proceedings of the conference held 16-21 October, 2006* by L.C. Ho, J,-M. Wang. vol. 373 (Xi'an Jioatong University, Xi'an, China, 2007), p. 267
42. E. Quataert, A. Gruzinov, ApJ **539**, 809 (2000)
43. E. Quataert, A. Gruzinov, ApJ **545**, 842 (2000)
44. A. Sadowski, R. Narayan, R. Penna, Y. Zhu, MNRAS **436**, 3856 (2013)
45. N.I. Shakura, R.A. Sunyaev, A&A **24**, 337 (1973)
46. J.M. Stone, J.E. Pringle, MNRAS **322**, 461 (2001)
47. J.M. Stone, J.E. Pringle, M.C. Begelman, MNRAS **310**, 1002 (1999)
48. M. Su, T.R. Slatyer, D.P. Finkbeiner, ApJ **724**, 1044 (2010)
49. M. Tahara, J. Kataoka, Y. Takeuchi et al., ApJ **802**, 91 (2015)
50. T. Totani, PASJ **58**, 965 (2006)
51. J.C. Vernaleo, C.S. Reynolds, ApJ **645**, 83 (2006)
52. Q.D. Wang et al., Science **341**, 981 (2013)
53. F.G. Xie, F. Yuan, MNRAS **427**, 1580 (2012)
54. F. Yuan, MNRAS **324**, 119 (2001)
55. F. Yuan, J. Lin, K. Wu, L. Ho, MNRAS **395**, 2183 (2009)
56. F. Yuan, D. Bu, M. Wu, ApJ **761**, 130 (2012)
57. F. Yuan, M. Wu, D. Bu, ApJ **761**, 129 (2012)
58. F. Yuan, R. Narayan, ARA&A **52**, 529 (2014)
59. F. Yuan, Z. Gan, R. Narayan, A. Sädowski, D. Bu, X. Bai, ApJ **804**, 101 (2015)
60. K. Zubovas, S. Nayakshin, MNRAS **424**, 666 (2012)
61. K. Zubovas, A.R. King, S. Nayakshin, MNRAS **415**, L21 (2011)

Chapter 5
A Brief Review of Relativistic Gravitational Collapse

Daniele Malafarina

Abstract We review here the basic setup to describe complete gravitational collapse of massive bodies within the general theory of relativity. We derive Einstein's equations describing collapse and solve them in some simple well-known toy models. We study the final outcome of collapse and the quantities that describe the formation of trapped surfaces and of the central singularity.

5.1 Introduction

General relativity became an essential part of the curricula of astrophysicists nearly 50 years after it was first developed by Albert Einstein, when new ultra-dense objects such as pulsars and highly energetic phenomena such as quasars were discovered. By 1963, it was clear that general relativity was necessary to understand those phenomena and that gravitational collapse played a crucial role for many astrophysical scenarios. Nevertheless, the seeds of our modern understanding of stellar collapse have deeper roots. It was Chandrasekhar, back in 1931, who used special relativity to evaluate the pressure necessary to overcome the electron degeneracy in a star and derived the famous upper mass limit for a stable white dwarf. He wrote: *"...the life history of a star of small mass must be essentially different from the life history of a star of large mass. For a star of small mass the natural white dwarf stage is an initial step towards complete extinction. A star of large mass cannot pass into the white dwarf stage, and one is left speculating on other possibilities"* [8]. The "other possibilities" to which Chandrasekhar referred today are called neutron stars and black holes.

D. Malafarina (✉)
Department of Physics and Center for Field Theory and Particle Physics,
Fudan University, 220 Handan Road, Shanghai 200433, China
e-mail: daniele.malafarina@nu.edu.kz

D. Malafarina
Physics Department, SST, Nazarbayev University, 53 Kabanbay Batyr Avenue,
Astana 010000, Kazakhstan

© Springer-Verlag Berlin Heidelberg 2016   169
C. Bambi (ed.), *Astrophysics of Black Holes*, Astrophysics
and Space Science Library 440, DOI 10.1007/978-3-662-52859-4_5

In 1939, Oppenheimer and Volkov performed a similar calculation using neutrons instead of electrons and general relativity instead of special relativity [39]. They concluded that any body with a large enough mass would not be able to sustain its own gravity and undergo complete collapse, although at the time it was not clear what the final state would be. They wrote: *"...actual stellar matter after the exhaustion of thermonuclear sources of energy will, if massive enough, contract indefinitely, although more and more slowly, never reaching equilibrium."* Soon after, Oppenheimer and Snyder and independently Datt developed the first exact solution of Einstein's equations describing a collapsing spherical cloud of non-interacting particles [11, 38]. The end state of such collapse model is a Schwarzschild black hole.

The existence of black holes as astrophysical objects was just a conjecture fifty years ago, while today is almost unanimously accepted by the community of astrophysicists. Nevertheless, the theoretical paradigm upon which the whole theory of black hole formation relies is not much different from the original Oppenheimer–Snyder–Datt (OSD) collapse model. And while our physical knowledge of astrophysical phenomena has progressed enormously in the last fifty years, our theoretical understanding of how black holes form is still very much rooted in simple toy models such as OSD. The reason for this relies mostly in the immense difficulty that one encounters when trying to solve analytically Einstein's equation in more general and physically relevant cases. Also, the fact that we do not know much about the behavior of matter in the strong field regime contributes in making our present theoretical understanding very limited.

On the other hand, from an experimental perspective, new missions and observatories are due to come online in the near future, and there is great hope that they will produce, among other things, an enormous amount of data on gravitational collapse and black hole formation. Astrophysicists of tomorrow will be able to rely on photons, neutrinos, and gravitational waves in order to study and understand what happens at the end of the life cycle of a star. This is usually called multimessenger astronomy. One of the key questions they will have to address is whether white dwarves, neutron stars, and black holes are the only possible objects that are left after a star dies. At present, it is natural to ask whether it is possible that there exists some yet unknown state of matter beyond the neutron degeneracy limit and capable of producing stable ultra-dense remnants. The purpose of this chapter is to pave the way for astrophysicists toward a broad theoretical understanding of the processes that lead to the formation of black holes. We do so by reviewing the paradigm for gravitational collapse in general relativity (GR) and the most fundamental analytical results that were obtained in the field.

The chapter is structured as follows: In Sect. 5.2, we derive the set of differential equations that are used to describe collapse. In Sect. 5.3, we discuss how the collapsing "star" can be matched to a vacuum exterior. Section 5.4 is devoted to the discussion of the conditions for the model to be physically viable. In Sect. 5.5, the apparent horizon and the singularity curve are defined. Section 5.6 presents the simplest solution for the homogenous dust collapse model, while in Sect. 5.7 inhomogeneous dust and homogeneous perfect fluid models are briefly outlined. Section 5.8 is devoted to discussing how the mathematical models can be useful for astrophysics, and finally, in Sect. 5.9, some possible future directions of investigation are discussed.

5.2   Einstein's Equations for the Collapsing Interior

Stars are supported in equilibrium by the balance of the gravitational attraction that pulls inward and the push outward generated by nuclear reactions happening at their center. When a star exhausts the nuclear fuel that was keeping it stable, it implodes under its own weight. At this point, the future evolution of the star depends on its mass. If a star is sufficiently massive, then there is no known force in nature capable of halting collapse. These stars end their lives forming black holes. In order to be able to describe the final stages of collapse, where the gravitational field becomes extremely large, we must use general relativity. Therefore, it is useful to begin our discussion by understanding what is a black hole in general relativity. The simplest and most intuitive definition of a black hole is that of a space-time singularity surrounded by an event horizon. Clearly, there are two elements that are crucial to our definition of a black hole: the singularity and the event horizon. The event horizon acts like a two-dimensional one-way membrane that lets particles and light enter while not letting anything exit. The singularity, strictly speaking, is not a part of the space-time, and it is the boundary that marks the geodesic incompleteness of all paths for particles that enter the horizon. The horizon for a non-rotating Schwarzschild black hole is located at a radius $R_{\text{Sch}} = 2GM_{\text{S}}/c^2$, where M_{S} is the black hole's mass, G is the Newton's constant, and c is the speed of light. The singularity is ideally "located" at the center of symmetry of the system. Intuitively, we can see that the black hole forms once enough mass is concentrated within a sphere of a small enough radius. Once matter is trapped inside the horizon, it can only fall toward the singularity (if there is no rotation). In principle, the equations of general relativity can be very difficult to solve; therefore, in order to describe the process by which all the matter in the star falls within the horizon radius thus forming a black hole, we need a mathematical framework that is simple enough to solve the equations but that still retains the most important physical features. In the following, we will neglect all the physical processes that happen in the cloud except gravity, we will assume that the cloud is perfectly spherically symmetric and not rotating, and we will assume that the exterior of the cloud is vacuum. Also, we shall consider here only extremely simplified fluid models to describe the state of matter of the collapsing star. Finally, it is custom to make use of natural units, thus setting $G = c = 1$.

5.2.1   Co-moving Coordinates

Our aim is to solve the system of Einstein's equations for a spherical collapsing matter cloud. As it is well known, Einstein's equations possess two distinct parts that both require some assumptions in order to allow us to find physically meaningful solutions. On the left-hand side, we have the geometrical part of the set of equations that is given by the Einstein tensor. This is determined once we know the metric for the space-time. As said, we will consider here only spherically symmetric, non-rotating space-times. A space-time is said to be spherically symmetric if the metric remains invariant under the group of spatial rotations $SO(3)$. This means that we can define the two-dimensional metric induced on the unit two-sphere as

$$d\Omega^2 = d\theta^2 + \sin^2\theta d\phi^2 \tag{5.1}$$

and define a function R for which $4\pi R^2$ represents the area of each two-sphere in the space-time. The full four-dimensional metric then can be written as

$$ds^2 = g_{AB}dx^A dx^B + R(x_A, x_B)^2 d\Omega^2, \tag{5.2}$$

with $A, B = 0, 1$. We can then introduce the coordinates $t = x_0$ and $r = x_1$ that diagonalize the two-dimensional part of the metric g_{AB} and write the most general spherically symmetric line element in the simple form

$$ds^2 = -e^{2\lambda}dt^2 + e^{2\psi}dr^2 + R^2 d\Omega^2 \tag{5.3}$$

where the functions λ, ψ, and R do not depend on the coordinates θ and ϕ. In the following, we will use a dot to express derivatives with respect to t and a prime to express derivatives with respect to r, thus writing $\dot{X} = dX/dt$, $X' = dX/dr$. The above coordinate system for which the metric is diagonal is called co-moving because one can think of the labels t and r as "attached" to each collapsing particle. Then, the functions λ, ψ, and R depend only on r and t. In this reference frame, the fluid is instantaneously at rest and its four-velocity u^μ is $u^t = e^{-\lambda}$, $u^r = u^\theta = u^\phi = 0$.

In order to describe the collapse of a spherically symmetric massive object such as a star, we need to solve Einstein's equations for a space-time described by (5.3) coupled to an energy momentum tensor describing a realistic fluid source. The energy momentum tensor is the right-hand side of Einstein's equations and for a fluid source in the co-moving frame can be written as

$$T^{\mu\nu} = \begin{pmatrix} \rho & 0 & 0 & 0 \\ 0 & p_r & 0 & 0 \\ 0 & 0 & p_\theta & 0 \\ 0 & 0 & 0 & p_\theta \end{pmatrix}$$

A perfect fluid is an idealized fluid where no shear stresses, no viscosity, and no heat conduction are present. It can be characterized by its mass density and isotropic pressures alone. Isotropic pressure means that the radial pressure equals the tangential pressure, and this implies $p_r = p_\theta = p$. The energy density is the energy per unit volume of the fluid in the local rest frame with four-velocity u^μ and the energy momentum tensor for a perfect fluid can then be written as

$$T^{\mu\nu} = (\rho + p)u^\mu u^\nu + pg^{\mu\nu}. \tag{5.4}$$

Note that in general relativity, the pressure contributes to the gravitational field, and therefore, the total gravitational energy need not be conserved. Nevertheless, the baryon number is conserved. For a gas of non-interacting particles, the so-called dust, the pressure vanishes and we can set $p = 0$. This is the simplest fluid model that can be considered.

5.2.2 Misner–Sharp Mass

The metric (5.3) can be used to describe static sources in equilibrium in the case when λ, ψ, and R do not depend on t. These are static objects with non-vanishing energy momentum. In this case, the area radius R can be used as a radial coordinate setting $R = r$. The simplest interior solution of this kind is given by the constant density Schwarzschild interior and was found by Schwarzschild himself together with the more famous vacuum solution (see, e.g., [47]). For the constant density interior, one sets $\rho = $ const. and solves Einstein's equations that take the form of the famous Tolman–Oppenheimer–Volkov equation [44], to find $p(r)$. Then, the object's boundary r_b is determined by the condition that $p(r_b) = 0$. For a metric describing a static interior case, we can define a function $m(r)$ such that

$$g_{rr} = e^{2\psi(r)} = \left(1 - \frac{2m(r)}{r}\right)^{-1}. \tag{5.5}$$

It is easy to see that at the boundary of the static object, the function $m(r)$ must become equal to the Schwarzschild parameter M_S that describes the total mass of the star, and therefore, we can interpret $m(r)$ as describing the amount of matter enclosed within the radius r. We can generalize the above expression in the dynamical case by introducing a function $U(r, t)$ as

$$U = u^\mu \frac{dR(r, t)}{dx^\mu} = e^{-\lambda}\dot{R}, \tag{5.6}$$

then, we get

$$g_{rr} = e^{2\psi(r,t)} = \left(1 + U^2 - \frac{2m(r, t)}{R}\right)^{-1} R'^2. \tag{5.7}$$

which reduces to the static case for $R = r$, so that $R' = 1$ and $\dot{R} = 0$. The Misner–Sharp mass $F(r, t)$ is then defined from $1 - F/R = g_{\mu\nu}\nabla^\mu R\nabla^\nu R$ and it is given by

$$F(r, t) = 2m(r, t) = R(1 - e^{-2\psi}R'^2 + e^{-2\lambda}\dot{R}^2). \tag{5.8}$$

In analogy with what was said before, we can interpret the Misner–Sharp mass as describing the amount of matter enclosed within the radius r at the time t [35].

5.2.3   Einstein's Equations

Einstein's equations couple the space-time geometry given by the metric $g_{\mu\nu}$ appearing in the Einstein's tensor $G_{\mu\nu}$ for the line element (5.3) to the matter content of the collapsing cloud given by the energy momentum tensor (5.4). Einstein's equations take the usual form

$$G_{\mu\nu} = R_{\mu\nu} - \frac{1}{2} g_{\mu\nu} R = T_{\mu\nu}, \tag{5.9}$$

where $R_{\mu\nu}$ and R are the Ricci tensor and Ricci scalar and where we have absorbed the constant factor $8\pi k$ into the definition of $T_{\mu\nu}$. Then, Einstein's tensor for the collapsing system is given by

$$G_t^t = -\frac{F'}{R^2 R'} + \frac{2\dot{R}e^{-2\lambda}}{RR'} \left(\dot{R}' - \dot{R}\lambda' - \dot{\psi}R' \right), \tag{5.10}$$

$$G_r^r = -\frac{\dot{F}}{R^2 \dot{R}} - \frac{2R'e^{-2\psi}}{R\dot{R}} \left(\dot{R}' - \dot{R}\lambda' - \dot{\psi}R' \right), \tag{5.11}$$

$$G_r^t = -e^{2\psi - 2\lambda} G_t^r = \frac{2e^{-2\lambda}}{R} \left(\dot{R}' - \dot{R}\lambda' - \dot{\psi}R' \right), \tag{5.12}$$

$$G_\theta^\theta = G_\phi^\phi = \frac{e^{-2\psi}}{R} \left((\lambda'' + \lambda'^2 - \lambda'\psi')R + R'' + R'\lambda' - R'\psi' \right) +$$
$$- \frac{e^{-2\lambda}}{R} \left((\ddot{\psi} + \dot{\psi}^2 - \dot{\lambda}\dot{\psi})R + \ddot{R} + \dot{R}\dot{\psi} - \dot{R}\dot{\lambda} \right). \tag{5.13}$$

These equations need to be supplemented with one more equation coming from the conservation of energy momentum that in general relativity comes as a consequence of the fact that the connection is metric and which can be written as

$$\nabla_\mu T_\nu^\mu = 0. \tag{5.14}$$

Then, in the simple case of pressureless (i.e., dust) collapse, and by making use of the definition of the Misner–Sharp mass given in Eq. (5.8), the first two equations of the above system simplify to

$$\rho = -G_t^t = \frac{F'}{R^2 R'}, \tag{5.15}$$

$$p = 0 = G_r^r = -\frac{\dot{F}}{R^2 \dot{R}}, \tag{5.16}$$

the third and fourth combine to give

$$\dot{R}' = \dot{R}\lambda' + \dot{\psi}R' = 0, \tag{5.17}$$

and the conservation of energy momentum (5.14) becomes

$$\rho\lambda' = 0. \tag{5.18}$$

From Eq. (5.16), we see that for dust, we must have $\dot{F} = 0$ which implies $F = F(r)$. This shows that the amount of matter enclosed within the co-moving radius r does not change with time. In other words, during collapse, there is no inflow or outflow of matter across any co-moving shell r. From Eq. (5.18), since the energy density is nonzero, we see that we must have $\lambda' = 0$, which implies $\lambda = \lambda(t)$. Now, we can define a new co-moving time coordinate $\tilde{t}$ by rescaling in such a way that

$$\frac{d\tilde{t}}{dt} = e^{\lambda}, \tag{5.19}$$

and therefore obtain

$$-e^{2\lambda}dt^2 = -e^{2\lambda}\left(\frac{dt}{d\tilde{t}}\right)^2 d\tilde{t}^2 = -d\tilde{t}^2. \tag{5.20}$$

This means that there is always the gauge freedom to fix the co-moving time t such that $\lambda = 0$, and in the following, we shall take t as such a gauge. Finally, Eq. (5.17) can be written as

$$\frac{\dot{R}'}{R'} = \dot{\psi}, \tag{5.21}$$

from which we get

$$R' = e^{g(r)+\psi}. \tag{5.22}$$

We call $f(r) = e^{2g(r)} - 1$ and the Misner–Sharp mass equation (5.8) can be rewritten in the form of the equation of motion of the system as

$$\dot{R}^2 = \frac{F(r)}{R} + f(r). \tag{5.23}$$

Once Eq. (5.23) is solved to give $R(r, t)$, the whole system of Einstein's equations is solved. The metric becomes

$$ds^2 = -dt^2 + \frac{R'^2}{1+f}dr^2 + R^2 d\Omega^2. \tag{5.24}$$

This is the well-known Lemaitre–Tolman–Bondi (LTB) space-time [6, 33, 43]. We see that the whole problem allows for two free functions of r, namely F and f, to be specified at will. As said, the function F can be thought of as representing the matter profile within the radius r, while from the above line element, the function f can be thought of as an energy profile describing the spatial curvature of the space-time. Then, provided that F and f are sufficiently regular, a unique solution of the equation of motion (5.23) exists for each regular initial condition $R_i = R(r, t_i)$.

5.3   Matching with an Exterior Metric

The metric given in Eq. (5.24) describes the dynamical collapse of a dust sphere. This can be thought of as describing the final stages of the life of a star, provided that gravity prevails on all other forces (thus allowing us to neglect any effect coming from the microphysics of the collapsing fluid) and that the collapsing cloud has a boundary. In the co-moving frame, this boundary can be identified with the surface given by the co-moving radius $r = r_b$. We consider the exterior of the collapsing dust cloud to be static and vacuum. Then, Birkhoff's theorem implies that it must be a portion of the Schwarzschild space-time. The exterior Schwarzschild solution can readily be derived from the metric (5.3) by including the further assumptions of staticity and vanishing of energy momentum tensor. A space-time is said to be stationary if it possess a time-like Killing vector, ∂_t. In such a case, the metric (5.3) becomes invariant under translations in t, and this is reflected in the metric functions λ, ψ, and R that do not depend on t. Further, a space-time is said to be static if the time-like Killing vector is orthogonal to the hypersurfaces of constant t. If we further impose that the energy momentum tensor is that of vacuum, namely $T_{\mu\nu} = 0$, we find that the only static spherically symmetric vacuum solution of Einstein's field equations can be written in the form

$$ds^2 = -\left(1 - \frac{2M_S}{r_s}\right) dt_s^2 + \left(1 - \frac{2M_S}{r_s}\right)^{-1} dr_s^2 + r_s^2 d\Omega^2. \qquad (5.25)$$

which is the well-known Schwarzschild line element expressed in Schwarzschild coordinates $\{t_s, r_s, \theta, \phi\}$. As said, the Schwarzschild metric has a singularity at the center $r_s = 0$. One way to determine the presence of singularities in solutions of Einstein's field equations is by inspecting curvature invariants looking for divergences. The Kretschmann scalar is one of these invariants and it is defined starting from the Riemann tensor as $\mathscr{K} = R_{\mu\nu\sigma\delta} R^{\mu\nu\sigma\delta}$. The Kretschmann scalar for Schwarzschild is

$$\mathscr{K} = 12\frac{4M_S^2}{r_s^6}, \qquad (5.26)$$

from which we see that the null surface $r_s = 2M_S$ is not singular; in fact, it is the event horizon [15], and the singularity is located at $r_s = 0$.

The global solution for the collapsing star is obtained by matching the collapsing interior given by the metric (5.24) to the vacuum exterior across a shrinking boundary surface Σ. Mathematically, "matching" means that the induced metric on the boundary hypersurface Σ must be the same on both sides. Also, the rate of change of the unit normal to Σ must be the same on both sides [23]. Let us label the interior metric with $(-)$ and the exterior metric with $(+)$. Then, the two line elements can be written as

$$ds_\pm^2 = g_{\mu\nu}^\pm dx_\pm^\mu dx_\pm^\nu. \tag{5.27}$$

The boundary hypersurface is implicitly defined on each side by $\Phi^\pm(x_\pm^\mu(y^a)) = 0$, and the induced metric on Σ can be written as

$$ds_\Sigma^2 = \gamma_{ab} dy^a dy^b, \tag{5.28}$$

where $a = 1, 2, 3$. Now define the three basis 4-vectors tangent to Σ as $e_{(a)}^\mu = \partial x^\mu / \partial y^a$. Then, the condition that the induced metric $\gamma_{ab}^\pm = g_{\mu\nu}^\pm e_{(a)}^\mu e_{(b)}^\nu$ agrees on both sides is simply $\gamma_{ab}^+ = \gamma_{ab}^-$. Define the unit normal to Σ as

$$n_\mu = \left(g^{\rho\sigma} \frac{\partial \Phi}{\partial x^\rho} \frac{\partial \Phi}{\partial x^\sigma} \right)^{-1/2} \frac{\partial \Phi}{\partial x^\mu}. \tag{5.29}$$

The extrinsic curvature (or second fundamental form) is defined as

$$K_{ab} = g_{\mu\nu} n^\mu \nabla_a e_{(b)}^\nu. \tag{5.30}$$

Then,

$$K_{ab}^\pm = \frac{\partial x_\pm^\mu}{\partial y^a} \frac{\partial x_\pm^\nu}{\partial y^b} \nabla_\mu n_\nu , \tag{5.31}$$

and continuity of K across Σ is given by

$$K_{ab}^+ = K_{ab}^-. \tag{5.32}$$

In the case of spherical collapse of dust, we have that the boundary hypersurface, in the exterior, with coordinates $\{x^+\} = \{t_s, r_s, \theta, \phi\}$, is given by

$$\Phi^+ = r_s - R_b(t_s) = 0, \tag{5.33}$$

and in the interior, with coordinates $\{x^-\} = \{t, r, \theta, \phi\}$, is given by

$$\Phi^- = r - r_b = 0. \tag{5.34}$$

So that the induced metric on the boundary is

$$ds_\Sigma^2 = -dt^2 + R_b(t)^2 d\Omega^2. \tag{5.35}$$

The Schwarzschild time t_s can be written as a function of the co-moving time t from

$$\frac{dt}{dt_s} = \sqrt{\left(1 - \frac{2M_S}{R_b}\right) - \left(1 - \frac{2M_S}{R_b}\right)^{-1}\left(\frac{dR_b}{dt_s}\right)^2}, \tag{5.36}$$

and the matching conditions for the continuity of the metric become

$$R_b(t_s) = R(r_b, t_s(t)), \tag{5.37}$$

$$F(r_b) = 2M_S. \tag{5.38}$$

Finally, continuity of K_{ab} follows identically from the matching conditions. Therefore, we see that the Misner–Sharp mass can be interpreted as the mass enclosed within the co-moving radius r and that on the boundary, it becomes proportional to the Schwarzschild mass M_S. Also, we see that the area function $R(r, t)$ in the interior at the boundary becomes the shrinking area radius in the Schwarzschild portion of the space-time. For more general collapse model, a matching to a suitable exterior space-time can also be defined (see, e.g., [13, 14, 27]).

5.4 Regularity, Scaling, and Energy Conditions

In order for the model to be physically acceptable, we need to choose an initial configuration that satisfies several conditions. The most important ones are regularity, which corresponds to requiring that the initial matter profiles do not present any singularities and are well behaved and the usual energy conditions, which in the dust case can be expressed via positivity of the energy density. Another requirement that is often imposed on the model is the absence of shell crossing singularities. These are caustics like singularities that are due to the overlap of infalling shells. Shell crossing singularities can possibly be removed by a suitable redefinition of the coordinates and generally do not represent a breakdown of the model.

5.4.1 Regularity and Scaling

We now investigate regularity of the matter profiles and the condition for avoidance of singularities at the initial time. In order to study these properties, we first need to express the area radius R, the Misner–Sharp F mass, and the velocity profile f in an appropriate gauge. As mentioned before, the Kretschmann scalar constitutes a valid tool to investigate the occurrence of singularities. In the case of the metric (5.24), this becomes

$$\mathscr{K} = 12\frac{F'^2}{R^4 R'^2} - 32\frac{F F'}{R^5 R'} + 48\frac{F^2}{R^6}. \tag{5.39}$$

We note here that with the present choice of the metric functions, one may be induced to think that the central curve $R = 0$ is always singular, including at the initial time. Nevertheless, this is not a physical singularity, as can be easily verified by evaluating the energy density, which turns out to be finite at the initial time. We notice then that there is a gauge degree of freedom in the scaling of R, namely in the way R is "measured" at the initial time that can be used to remove the above ambiguity. In fact, we can always choose arbitrarily the initial value of R. In the following, we choose this initial scaling condition as

$$R(r, t_i) = r. \tag{5.40}$$

From the choice of the initial data for R given in Eq. (5.40), we see that the gauge freedom allows us to define a scaling function $a(r, t)$ from the area function $R(r, t)$ as

$$R = r a(r, t). \tag{5.41}$$

Now, the scaling factor a is an a-dimensional quantity such that

- at the initial time, we have $a(r, t_i) = 1$,
- at the time of formation of the singularity t_{sing}, we have $a(r, t_{sing}) = 0$,
- collapse is given by $\dot{a} < 0$.

For dust, using Einstein's equation (5.15), the above scaling implies that the initial density must satisfy the following condition:

$$\rho(r, t_i) = \rho_i(r) = \frac{F'}{r^2} > 0. \tag{5.42}$$

Therefore, in order to avoid having ρ diverging at $r = 0$ at the initial time, we must impose a regularity condition on the Misner–Sharp mass. This is given by

$$F(r) = r^3 M(r), \tag{5.43}$$

with $M(r)$ non-diverging and sufficiently regular in the interval $[0, r_b]$. Generally, we assume that the function $M(r)$ can be written as a polynomial expansion in the vicinity of $r = 0$. In general, a physically viable density profile should be non-increasing radially outwards and therefore we must impose that the first non-vanishing term in the polynomial expansion of M is vanishing or negative in $r = 0$. It is reasonable to suppose that $M' \le 0$ near the center. With the above scaling, the density becomes

$$\rho = \frac{3M + rM'}{a^2(a + ra')}.$$ (5.44)

If we add the further requirement that ρ must not present any cusps at $r = 0$, we must impose that $M'(r)$ vanishes in $r = 0$, and therefore, we must require $M'' \leq 0$ near $r = 0$. Note that in the simplest case of homogeneous dust, ρ does not depend on r and so $M(r)$ must be constant M_0. Then,

$$\rho(t) = \frac{3M_0}{a^3},$$ (5.45)

and as a consequence, the scale factor also does not depend on r. With this choice of the scaling factor, it is easy to verify that the central density diverges only at the singularity. Also, we see that the Kretschmann scalar in the homogeneous case reduces to

$$\mathscr{K} = 60\frac{M_0^2}{a^6},$$ (5.46)

and it is regular at the initial time, its value being $\mathscr{K}_i = 60M_0^2$. In general, for inhomogeneous dust, we have

$$\mathscr{K} = 12\frac{(3M + rM')^2}{a^4(a + ra')^2} - 32\frac{M(3M + rM')}{a^5(a + ra')} + 48\frac{M^2}{a^6}.$$ (5.47)

Note that by writing $\mathscr{K}$ in terms of M and a, we avoid the problem of divergence along the central line. In the new scaling along $r = 0$, we see that $\mathscr{K}$ diverges only for $a = 0$, thus showing the occurrence of the singularity. The curve $t_{sing}(r)$ for which $a(r, t_{sing}) = 0$ is the singularity curve which describes the time at which the shell r becomes singular. As a consequence of the fact that in the homogeneous case, a depends only on t, we see that for homogeneous dust, the singularity occurs at the same time t_{sing} for every co-moving shell r.

From the equation of motion (5.23), we see that at the initial time, the velocity of the infalling particles is given by

$$\dot{R}_i = -\sqrt{\frac{F}{r} + f}.$$ (5.48)

Given the fact that the choice of the free function F corresponds to fixing the initial density profile from the above equation, we see that fixing f corresponds to determining the initial velocity profile for the particles in the cloud. Now, by making use of the scaling above, we can rewrite Eq. (5.48) as

$$\dot{a}_i = -\sqrt{M + \frac{f}{r^2}},$$ (5.49)

from which we see that in order to have a finite initial velocity at all radii, we must set a scaling for f as well. We shall take

$$f(r) = r^2 b(r), \qquad (5.50)$$

with $b(r)$ a sufficiently regular function (again which can be given as a polynomial expansion near $r = 0$). To summarize, at the initial time, we have the freedom to specify three functions of r as follows:

- Choose an initial condition for the scaling $R(r, t_i) = R_i(r)$ or equivalently set the value of $a(r, t_i)$.
- Choose a mass function $F(r)$, or equivalently $M(r)$, which implies an initial density $\rho_i = F'/r^2$.
- Choose a velocity function $f(r)$, or equivalently $b(r)$, which implies the initial condition for the velocity $\dot{R}(r, t_i)$.

Then, the system is fully determined and the equation of motion can be written as

$$\dot{a}(r, t) = -\sqrt{\frac{M(r)}{a(r, t)} + b(r)}. \qquad (5.51)$$

By solving the above equation for a, we completely solve the system of Einstein's equations. As said, homogeneous dust collapse is given by $\rho = \rho(t) = 3M_0/a^3$. Therefore, from the above, it follows that homogeneous dust collapse can be obtained from the following requirements:

- $a = a(t)$
- $M(r) = M_0 = \text{const.}$
- $b(r) = k = \text{const.}$

In this case, we can give a precise interpretation of the velocity profile if we imagine a dust cloud that extends to infinity. We can think at the constant k as representing the initial velocity of particles at spatial infinity, and we can characterize the geometry of the space-time based on the sign of k in the following way:

- $k = 0$ marginally bound collapse, corresponding to a flat geometry. Shells at radial infinity begin collapse with zero initial velocity.
- $k > 0$ unbound collapse, corresponding to a hyperbolic geometry. Shells at radial infinity have positive initial velocity.
- $k < 0$ bound collapse, corresponding to an elliptic geometry. Shells at radial infinity have negative initial velocity.

Note that if one wishes to have zero initial velocity $\dot{R}_i = 0$ for particles in the collapsing cloud with boundary, then the only possible choice is that of bound collapse with $M_0 = -k$.

5.4.2 Energy Conditions

Einstein's equations are often regarded as made of two different parts. The "golden" half, more elegant, is the left-hand side that contains the Einstein tensor and therefore encodes the information about the geometry of the space-time. The "wooden" half is the right-hand side that contains the energy momentum tensor and in principle should describe the physical properties of matter. As a matter of fact, it is generally practically impossible to fully describe all the properties of the matter fields in the energy momentum tensor, and therefore, one usually resorts to simplifications and averaged properties that are valid for macroscopic fields. Nevertheless, we must keep in mind that the behavior of matter under very strong gravitational fields is not known at present, and therefore, the description of classical macroscopic fluids that is valid in the weak field may not be enough when the curvature becomes very high. To simplify things, one usually imposes that certain inequalities be satisfied by the energy momentum tensor in order for the same to be considered physically viable [18]. The first and most commonly used inequality is the weak energy condition (w.e.c.). To satisfy the w.e.c., the energy momentum tensor must be given in such a way that $T_{\mu\nu}V^\mu V^\nu \geq 0$ for any time-like (and null) vector V^μ. This means that the energy density must be nonnegative in any reference frame. The energy momentum tensor for a fluid made of massive particles, with respect to some orthonormal basis, can always be written as $T^{\mu\nu} = diag\{\rho, p_1, p_2, p_3\}$. Then, the weak energy conditions in the co-moving frame can be written as

$$\rho \geq 0 \quad \rho + p_i \geq 0 \quad \text{with} \quad i = 1, 2, 3. \tag{5.52}$$

This is the less demanding of the energy requirements. The weak energy condition allows for violations of the conservation of baryon number as new particles can be created. Still, more stringent conditions can be imposed. If one desires to impose that the total amount of mass in the space-time is conserved, then one must impose the dominant energy condition (d.e.c.) which states that for every time-like vector V^μ, the energy momentum tensor must satisfy both $T_{\mu\nu}V^\mu V^\nu \geq 0$ and $T_{\mu\nu}V^\mu$ being a null or time-like vector. This means that not only the energy density is nonnegative in any frame, but also the flow of ρ must be locally not space-like. As a consequence, we get that in an orthonormal reference frame, the energy density must be greater than the pressures. Namely,

$$\rho \geq 0 , \quad -\rho \leq p_i \leq \rho \quad \text{with} \quad i = 1, 2, 3. \tag{5.53}$$

Note that if we define the speed of sound waves within the fluid travelling in the direction of p_i as $v_i = dp_i/d\rho$ $(i = 1, 2, 3)$, then the d.e.c. does not allow for the speed of sound to be greater that the speed of light. It is a very reasonable assumption that is not implemented by the w.e.c.. A fluid that satisfies the d.e.c. obviously satisfies also the w.e.c.. Finally, let us briefly mention a third energy condition that can be imposed and that is not directly related to the previous two. This is the strong energy condition (s.e.c.), and for a perfect fluid in the co-moving frame, it is equivalent to requiring

$$\rho \geq 0 \, , \ \rho + p \geq 0 \, , \ \rho + 3p \geq 0. \tag{5.54}$$

In the following, we shall always require that the fluid satisfies the w.e.c. and whenever possible that it satisfies the d.e.c. as well.

5.4.3   Shell Crossing Singularities

From Eq. (5.47), we see that the Kretschmann scalar diverges when the central singularity forms, namely when $a = 0$, but also when $R' = 0$ if $M' \neq 0$. In this case, we speak of the occurrence of shell crossing singularities. These are true curvature singularities that arise from overlapping radial shells. At the shell crossing singularity, the radial geodesic distance between shells with radial coordinate r and $r + dr$ vanishes. These singularities are equivalent to caustics in wave propagation, and it is reasonable to assume that the space-time can be extended through the singularity by a suitable redefinition of the coordinates. This can also be seen from the fact that shell crossing singularities are gravitationally weak, meaning that geodesics reaching the singularity are not squeezed into a line (as is the case of the central singularity), and thus, observers at the shell crossing are not crushed (see, e.g., [21, 22, 32, 49, 50]). Nevertheless, in any collapse model, a condition that can be required is the absence of shell crossing singularities. In order to avoid shell crossing singularities, we can either impose $R' \neq 0$ or choose $M(r)$ in such a way that $M(r) = 0$ when $R' = 0$ so that $M'/R' < \infty$. During collapse, the mass function M is generally assumed to be positive; therefore, requiring the absence of shell crossing singularities is equivalent to requiring that R' is not vanishing. Note that since $R' = a + ra'$ in a neighborhood of the center, the condition can always be satisfied if $a' \neq 0$.

5.5   Trapped Surfaces and Singularities

In the Schwarzschild space-time, the surface $r_s = 2M_S$, known as the event horizon, is *"...a perfect unidirectional membrane: causal influences can cross it in only one direction"* [15]. The event horizon is the boundary of the region where light rays can not escape to infinity. At the horizon, the time-like Killing vector is null and outgoing null geodesics have zero radial velocity. Nevertheless, the event horizon is not a very useful concept for practical (i.e., astrophysical) purposes. In fact, the event horizon is a global property of the space-time which does not depend on the observer and its determination requires the knowledge of the entire future history of the space-time. What we need in order to be able to make experiments is a local approach to the definition of trapped surfaces that allows us to define when a co-moving observer that is collapsing with the cloud becomes causally disconnected from the outside universe. If matter is present, as in the case of the LTB metric given in Eq. (5.24), the event horizon is not the only possible horizon that can be defined and it is not the most useful concept to investigate the physics that occurs as the black hole forms. We want to know how and when the black hole forms during gravitational collapse. When light will be trapped by the gravitational field? The exterior region will become a black hole solution once the boundary surface $R_b(t)$ passes the Schwarzschild radius. What about the interior? Each collapsing shell will become causally disconnected from the outer universe at some point and will eventually fall into the central singularity. If we want to track the formation of the horizon inside the matter region, first we need to know what we mean by trapped surface in the interior.

Given a 3 + 1 slicing of the space-time, consider the three-dimensional space-like slice. Then, a "trapped surface" is defined as a smooth closed two-surface in the slice whose future-pointing outgoing null geodesics have negative expansion. This means that all light rays, all null geodesics, emanating from the surface are pointing inward. The "trapped region" in the slice is then defined as the union of all trapped surfaces, and the "apparent horizon" is the outer boundary of the trapped region [7, 20, 40]. One intuitive way to understand the difference between the apparent horizon and the event horizon is to note that the event horizon is the surface at which any light ray directed outward can be initially outgoing and eventually become ingoing, thus falling back inwards at some later time, while the apparent horizon is the surface for which all light rays directed outwards are ingoing, thus directed inward at the time when they are emitted. In vacuum, the two surfaces coincide, and therefore, the apparent horizon and the event horizon in the Schwarzschild space-time are the same. Still, when matter is present, they can be different, as is the case for the LTB metric.

The apparent horizon in general need not be a null surface and it always lies inside the event horizon. Nevertheless, it is the apparent horizon that determines the trapped region in the collapsing cloud. It is a local property of the space-time and is observer dependent, and therefore, it can be experimentally tested, while the event horizon may be undetectable. To understand this, imagine the situation of a thin spherical shell separating a vacuum Minkowski interior from a vacuum Schwarzschild exterior. The shell may be time-like or light-like. Let the shell have total mass M_S and collapse under its own gravity (see Fig. 5.1). As the shell collapses, an event horizon will form at $R = 0$ at the time $t = t_0$. The event horizon curve will expand to larger radii and eventually match the Schwarzschild radius $R = 2M_S$ in the exterior at $t = t_{\text{Sch}}$. An observer living at a fixed radial coordinate $R = R_1$ inside the Minkowski region will experience the event horizon passing through him at the time t_1 but will not have any way to detect it. This shows how in principle we can have event horizons where we would not expect and why local experiment cannot detect the presence of an event horizon. For this reason, the apparent horizon is a more useful tool to study the trapped region that develops during the formation of a black hole.

For the spherical dust collapse model, the apparent horizon is the surface for which the surface $R(r, t) = $ const. is null. This means

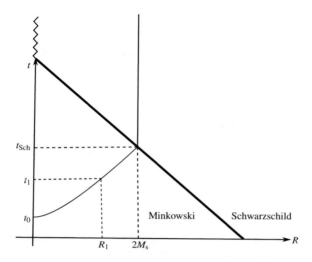

Fig. 5.1 Collapse of a thin spherical null shell (*thick line*) separating a flat vacuum interior from a Schwarzschild exterior. The angular coordinates θ and ϕ are suppressed, so every radius R corresponds to a spherical surface. The *thin line* represents the event horizon, which forms at the center of symmetry of the system at the time t_0 and expands toward bigger radii. As the shell crosses the Schwarzschild radius $2M_S$, the horizon settles to the usual event horizon of a static black hole. Observers living in the interior at a fixed radius R_1 would fall inside the trapped region at the time t_1 but would have no way of detecting the horizon that is passing through them

$$g^{\mu\nu}(\partial_\mu R)(\partial_\nu R) = 1 + f - e^{-2\lambda}\dot{R}^2 = 0. \qquad (5.55)$$

From the definition of the Misner–Sharp mass in Eq. (5.8), we get that the condition for the formation of trapped surfaces can be expressed as

$$1 - \frac{F}{R} = 1 - \frac{r^2 M}{a} = 0. \qquad (5.56)$$

This condition can be viewed as the implicit definition of the curve $t_{ah}(r)$ for which

$$a(r, t_{ah}(r)) = r^2 M(r). \qquad (5.57)$$

The above curve is called the apparent horizon curve and describes the co-moving time t at which the co-moving shell r becomes trapped.

As we have seen, the other important curve to describe the formation of the black hole at the end of collapse is the singularity curve $t_{sing}(r)$. This is the curve that describes the co-moving time t at which the co-moving shell r becomes singular. This curve represents the limit of the space-time manifold, and all geodesics inside the trapped region must terminate at the singularity. From the condition of formation of the singularity, we see that the curve $t_{sing}(r)$ is given implicitly by

$$a(r, t_{sing}(r)) = 0. \tag{5.58}$$

We have already seen that in the case of homogeneous dust, we must have $t_{sing} =$ const., which means that all shells become singular at the same co-moving time.

5.6 Homogeneous Solutions

The equation of motion (5.51) for homogeneous collapse is written as

$$\dot{a} = -\sqrt{\frac{M_0}{a} + k}. \tag{5.59}$$

We can characterize the geometry depending on the sign of the free parameter k by introducing the following change of coordinates

$$r = S_k(\chi) = \begin{cases} \sinh \chi & \text{if } k = +1, \text{ "hyperbolic" region,} \\ \chi & \text{if } k = 0, \text{ "flat" region,} \\ \sin \chi & \text{if } k = -1, \text{ "elliptic" region.} \end{cases}$$

We can then write the Oppenheimer–Snyder metric in a unified form as

$$ds^2 = -dt^2 + a(t)^2 \left[d\chi^2 + S_k(\chi)^2 d\Omega^2 \right], \tag{5.60}$$

and the solution of the equation of motion is given in parametric form by

$$a(t) = \begin{cases} \frac{M_0}{2k}(\cosh \eta - 1) & \text{with} \quad \sinh \eta - \eta = \frac{2k^{3/2}(t-t_s)}{M_0} & \text{if } k > 0, \\ \left(\frac{3M_0(t-t_s)}{2} \right)^{2/3} & & \text{if } k = 0, \\ \frac{M_0}{-2k}(1 - \cos \eta) & \text{with} \quad \eta - \sin \eta = \frac{2(-k)^{3/2}(t-t_s)}{M_0} & \text{if } k < 0, \end{cases}$$

On the other hand, one can always solve the equation of motion (5.59) to find $t(a)$.

• In the flat region given by $k = 0$, the equation of motion is easily integrated to give

$$t(a) = -\frac{2a^{\frac{3}{2}}}{3\sqrt{M_0}} + t_{sing}.$$

• In the hyperbolic region, given by $k > 0$, we define $X = M_0/k$ and we get

$$t(a) = \frac{a}{\sqrt{k}} \left(\frac{X}{a} \tanh^{-1} \frac{1}{\sqrt{\frac{X}{a} + 1}} - \sqrt{\frac{X}{a} + 1} \right) + t_{sing}.$$

• In the elliptic region, given by $k < 0$, we define $X = -M_0/k$ and we get

$$t(a) = \frac{a}{\sqrt{-k}} \left(\sqrt{\frac{X}{a} - 1} - \frac{X}{a} \tan^{-1} \frac{1}{\sqrt{\frac{X}{a} - 1}} \right) + t_{sing}.$$

Finally, for homogeneous dust, the metric (5.3) can also be written as

$$ds^2 = -dt^2 + \frac{a^2}{1 + r^2 k} dr^2 + r^2 a^2 d\Omega^2, \tag{5.61}$$

from which we see that the metric in the interior is just the time reversal of a dust Friedmann–Robertson–Walker cosmological solution.

Marginally bound case is a particularly simple case given by $k = 0$. Then, the initial velocity of collapse is nonzero, as can be seen from $\dot{a}_i = -\sqrt{M_0}$, and the equation of motion is simply $\dot{a} = -\sqrt{M_0/a}$ for which the solution is obtained immediately as (see Fig. 5.2)

$$a(t) = \left(1 - \frac{3}{2}\sqrt{M_0} t \right)^{2/3}. \tag{5.62}$$

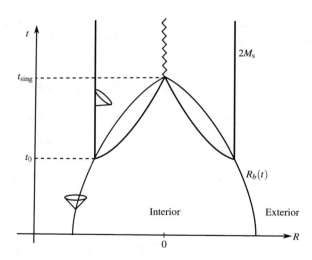

Fig. 5.2 Schematic view of homogeneous dust collapse. At the time $t_i = 0$, no singularities are present. The boundary curve $R_b(t)$ separates the interior from the vacuum exterior. The cloud collapses as t increases, and at the time t_0, the horizon forms at the boundary. In the exterior region, the horizon is the Schwarzschild radius. In the interior, the apparent horizon propagates inward and reaches the center of symmetry at the time of formation of the singularity t_{sing}. For $t > t_{sing}$, the space-time has settled to the usual Schwarzschild solution

5.6.1 Apparent Horizon and Singularity

From the condition for formation of trapped surfaces given in Eq. (5.56), we obtain the curve $t_{ah}(r)$ describing the time at which the shell r crosses the apparent horizon:

$$
t_{ah}(r) = \begin{cases}
t_{sing}(r) + \frac{F}{f^{\frac{3}{2}}} \tanh^{-1}\sqrt{\frac{f}{1+f}} - \frac{F}{f}\sqrt{1+f}\,, & \text{for } k > 0, \\[2mm]
t_{sing}(r) - \frac{2}{3}F(r) = \frac{2}{3\sqrt{M_0}} - \frac{2}{3}r^3 M_0\,, & \text{for } k = 0, \\[2mm]
t_{sing}(r) + \frac{F}{(-f)^{\frac{3}{2}}} \tan^{-1}\sqrt{-\frac{f}{1+f}} - \frac{F}{f}\sqrt{1+f}\,, & \text{for } k > 0.
\end{cases}
$$

Now, if we look for simplicity at the apparent horizon for marginally bound homogeneous dust model, we see that it forms initially at the boundary of the collapsing cloud at the time $t_0 = t_{ah}(r_b) = 2/3\sqrt{M_0} - 2r_b^3 M_0/3$ and then propagates inward toward the center. The time t_0 is the same time at which the event horizon forms in the exterior spacetime. For $t > t_0$, the apparent horizon curve moves to smaller radii reaching the center at the time $t_{sing} = t_{ah}(0) = 2/3\sqrt{M_0}$, which is the time of formation of the singularity. Inside the trapped region, all geodesics terminate at the singularity; therefore, an observer on the boundary, once this has passed the horizon, falls toward the singularity in a finite time of the order of $\sqrt{R_b^3/GM}$. An observer at infinity sees the boundary approaching the horizon becoming infinitely redshifted and indefinitely slow [36].

The singularity is reached once the density diverges. From the above, we have seen that the shell focusing strong curvature singularity corresponds to $a = 0$. In the homogeneous dust collapse case, this gives

$$
t_{sing} = \begin{cases}
\frac{1}{\sqrt{-k}}\left(X\tan^{-1}\frac{1}{\sqrt{X-1}} - \sqrt{X-1}\right)\,, & \text{for } k > 0, \\[2mm]
\frac{2r^{\frac{3}{2}}}{3\sqrt{F}} = \frac{2}{3\sqrt{M_0}}\,, & \text{for } k = 0, \\[2mm]
\frac{1}{\sqrt{k}}\left(\sqrt{X+1} - X\tanh^{-1}\frac{1}{\sqrt{X+1}}\right)\,, & \text{for } k > 0.
\end{cases}
$$

The singularity is simultaneous, and all shells fall into the singularity at the same co-moving time.

5.7 Inhomogeneous Dust and Collapse with Pressures

The easiest way to extend the homogeneous dust collapse model is to introduce inhomogeneities in ρ. Inhomogeneous models have been widely considered in cosmology which are obtained by a time reversal of collapse models (see [5, 31] and references therein). This means considering $\rho = \rho(r, t)$, with ρ radially decreasing in order for the matter profile to be physically realistic. This describes a dust cloud that initially has higher density at the center. Following the same procedure as in homogeneous collapse, we can evaluate Einstein's equations as

$$\rho = \frac{F'}{R^2 R'}, \tag{5.63}$$

$$\dot{F} = 0, \tag{5.64}$$

$$\lambda' = 0, \tag{5.65}$$

$$\dot{\psi} = \frac{\dot{R}'}{R'}. \tag{5.66}$$

The main difference with the homogeneous case is that now we will have $M(r)$, $b(r)$, and $a(r, t)$. Then, it is worth asking how the boundary, the trapped surfaces, and singularity are affected by the presence of inhomogeneities. Does a black hole still form at the end of collapse? Yes, the singularity theorems by Hawking and Penrose tell us that once the trapped surfaces form, the formation of the singularity is inevitable. Eventually, all matter falls into the central singularity and we are left with a Schwarzschild black hole [19]. Nevertheless, if we ask whether we get a picture of collapse qualitatively similar to the OS model, then the answer is not always in the affirmative. In fact, the way in which the apparent horizon and the singularity curve develop depends on the form of the density and velocity profiles and some important differences with the OS model may arise. The most striking of these differences is that in the inhomogeneous dust collapse, there is the possibility for the central singularity to be "naked" (i.e., not covered by a horizon) at the instant of formation (see, e.g., [9, 12, 26, 37, 48]).

In general, if $M = M(r)$, we have

$$\rho = \frac{F'}{R^2 R'} = \frac{3M + rM'}{a^2(a + ra')}. \tag{5.67}$$

If we consider $M(r)$ as a polynomial expansion near $r = 0$, we can take

$$M(r) = M_0 + M_1 r + M_2 r^2 + \cdots, \tag{5.68}$$

and the condition for the energy density to be radially decreasing outward is given by $M_1 \leq 0$. If we also wish to impose that the density does not present cusps at the center, we may impose $M_1 = 0$, and then, the condition for ρ to be decreasing becomes $M_2 \leq 0$. This is consistent with the choice of density profiles in astrophysical models that generally present only quadratic terms in r. Similar to the homogeneous case, the value $a = 0$ signals the appearance of the shell focusing singularity. On the other hand, now we need to make sure that shell crossing singularities, given by $R' = a + ra' = 0$, do not occur before the formation of the singularity. For simplicity, let us consider the marginally bound case. Then, $R' = 0$ implies

$$1 - \frac{3}{2}\sqrt{M} - \frac{rtM'}{2\sqrt{M}} = 0, \qquad (5.69)$$

which can be used to obtain the time at which the shell r develops a shell crossing singularity as

$$t_{sc}(r) = \frac{2\sqrt{M}}{3M + rM'}. \qquad (5.70)$$

For homogeneous dust, then $t_{sc} = t_{sing}$ and we do not have shell crossing singularities during collapse. Similarly, in the inhomogeneous case with $M' < 0$, we have $t_{sc}(r) \geq t_s(r)$, and therefore, no shell crossing singularities occur before the formation of the central singularity. From this, we see that the physical requirement of a radially decreasing density profile is compatible with the condition for avoidance of shell crossing singularities.

The solution for inhomogeneous dust collapse can be obtained form the one for homogeneous dust by replacing M_0 and k with $M(r)$ and $b(r)$. Again, let us consider for simplicity the solution for marginally bound collapse. This is given by

$$a(r, t) = \left(1 - \frac{3}{2}\sqrt{M(r)}t\right)^{2/3}. \qquad (5.71)$$

We immediately see that now each shell collapses with a different scale factor and a different velocity. As a consequence, each shell becomes singular at a different time. The apparent horizon curve is also affected, as now it does not necessarily form initially at the boundary. The singularity curve and apparent horizon curve are explicitly given by

$$t_{sing}(r) = \frac{2}{3\sqrt{M(r)}}, \qquad (5.72)$$

$$t_{ah}(r) = t_{sing}(r) - \frac{2}{3}r^3 M, \qquad (5.73)$$

and near $r = 0$, they have the same behavior up until the third order in r,

$$t_{sing}(r) = \frac{2}{3\sqrt{M_0}} - \frac{M_1 r}{3M_0^{3/2}} + \cdots, \qquad (5.74)$$

$$t_{ah}(r) = \frac{2}{3\sqrt{M_0}} - \frac{M_1 r}{3M_0^{3/2}} + \cdots. \qquad (5.75)$$

Note that near the center, the apparent horizon curve is increasing and the central line $r = 0$ becomes singular and trapped at the same time. This suggests the possibility for the existence of geodesics that originate at the central singularity and are not trapped inside the horizon as (see Fig. 5.3). Due to the lack of pressures in the model,

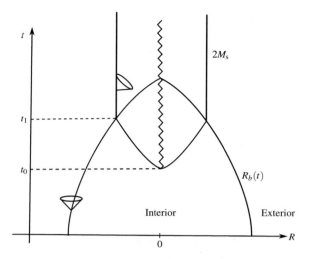

Fig. 5.3 Schematic view of inhomogeneous dust collapse. At the time $t_i = 0$, no singularities are present. The boundary curve $R_b(t)$ separates the interior from the vacuum exterior. The cloud collapses as t increases, and at the time t_0, the horizon forms at the center of the cloud. The singularity forms at the same time. Null geodesic can originate from the singularity and reach distant observers. In the interior, the apparent horizon propagates outward and reaches the boundary at the time $t_1 > t_0$. Once all the matter falls into the singularity, the space-time settles to the usual Schwarzschild solution

the boundary of the star r_b can be chosen arbitrarily. This is a mathematical artifact of the dust solution, and in the case with pressures, the boundary would have to be set at the radius where p vanishes. Therefore, if the boundary is chosen in such a way that t_{ah} is always increasing, it is possible to find null geodesics that originate at $t_s(0)$ and reach observers at infinity. Outgoing radial null geodesics $t_\gamma(r)$ are given by

$$\frac{dt_\gamma}{dr} = R', \qquad (5.76)$$

and it can be proven that there are null geodesics coming out of the first instant of the central singularity $t_s(0)$ and reaching the boundary [9, 12, 26, 37, 48]. Naked singularities are found in many solutions of Einstein's equations and can be very different from one another. The question is if they can form from physically realistic processes. How much these models rely on the assumptions? What outcome will come from more realistic models? We shall shortly discuss these issues in the next sections (for a more detailed discussion, see, e.g., [28]).

Analytically, we cannot deal with rotation or departures from spherical symmetry. In fact, when it comes to rotation at present, we do not possess an analytical solution that describes an interior for the Kerr space-time that matches smoothly to the exterior. Nevertheless, some indications on how general is the scenario described in the dust collapse model can be obtained by considering collapse of a fluid source with pressures. For a fluid source, the energy momentum tensor is given by $T_{\mu\nu} = diag\{\rho, p_r, p_\theta, p_\theta\}$ and we can write Einstein's equations and energy momentum conservation

$$\rho = \frac{F'}{R^2 R'}, \tag{5.77}$$

$$p_r = -\frac{\dot{F}}{R^2 \dot{R}}, \tag{5.78}$$

$$\lambda' = 2\frac{p_\theta - p_r}{\rho + p_r}\frac{R'}{R} - \frac{p_r'}{\rho + p_r}, \tag{5.79}$$

$$\dot{G} = 2\lambda'\frac{\dot{R}}{R}G, \quad \text{with} \quad G = R'^2 e^{-2\psi}. \tag{5.80}$$

Then, it is easy to see that the system of equations to be solved becomes much more complicated with respect to the dust case. Together with the Misner–Sharp mass definition, the system has five equations and seven unknown functions. Specifying equations of state for p_r and p_θ then closes the system.

In this case, the Misner–Sharp mass need not be conserved during the evolution, and therefore, there can be an inflow or outflow of matter across each shell r as collapse progresses. As a consequence, the matching with the exterior space-time need not be done with the Schwarzschild solution. Requiring the boundary condition $p_r(r_b, t) = 0$ implies that $F(r_b, t)$ is conserved during collapse and the exterior is Schwarzschild, and on the other hand, the condition that p_r vanishes at the boundary translates in a variable boundary surface $r_b(t)$. It can be shown that matching with the Vaidya solution describing ingoing or outgoing null dust can be done in certain cases and matching with a generalized Vaidya solution is always possible [13, 14, 27].

The simplest model with pressures that can be considered is that of a homogeneous perfect fluid with linear equation of state. Then, requiring that the fluid be perfect implies $p_r = p_\theta = p$, while homogeneity implies $p = p(t)$ and $\rho = \rho(t)$. Finally, a linear equation of state relates p to ρ via

$$p = \gamma\rho, \tag{5.81}$$

with γ being a constant. The presence of the linear equation of state closes the system, and Einstein's equations can be fully integrated in this case. The third Einstein's equation (5.79) becomes again $\lambda' = 0$, and the fourth equation gives again $G = 1 + kr^2$. Then, the equation of motion is again written as

$$\dot{a}^2 = \frac{M}{a} + k, \tag{5.82}$$

where now $M = M(t)$ is to be determined from the equation of state that together with Eqs. (5.77) and (5.78) gives

$$\dot{M} = -\frac{3\gamma M}{a}. \tag{5.83}$$

Integrating the above equation, we get

$$M(t) = \frac{M_0}{a^{3\gamma}}. \tag{5.84}$$

Einstein's equations (5.77) and (5.78) imply that the density is $\rho = 3M/a^3$ and the pressure is $p = -\dot{M}/a^2$. To have a positive pressure, M must decrease in time. If we require a constant co-moving boundary $r = r_b$, then the exterior metric cannot be a portion of the Schwarzschild space-time and some mass must be radiated away in the exterior region. This can be easily seen from the fact that $M(t)$ implies that the total mass within r_b changes with time, and therefore, there must be an outflow of matter from the co-moving boundary. As said, we can always match to a non-vacuum solution describing a radiating null fluid or we can require the matching to be performed at a surface $r_b(t)$.

5.8   Collapse in Astrophysics

We studied here some simple analytical toy models that describe the complete grav-itational collapse of a spherical matter cloud made of non-interacting particles in general relativity (GR). If we assume as a first approximation that these models can be used to describe the most relevant features of the collapse of the core of a massive star, we see that a black hole must inevitably form as the final product of collapse. At the time t_0, the horizon forms as the boundary of the star crosses the threshold of the Schwarzschild radius, and at the time $t_{sing} > t_0$, the singularity forms at the center of symmetry of the system. In the end, we are left with a Schwarzschild black hole. The Oppenheimer–Snyder–Datt (OSD) model is very simple and relies on many simplifying assumptions that while on the one hand allow us to solve the equations analytically thus finding a global solution, on the other hand make for a scenario that is not very realistic. The OSD model can be seen as the bridge between mathematical black holes and astrophysical black holes in the sense that it is a simplified mathe-matical description of a dynamical phenomena that nevertheless captures the most essential features. The main assumptions in the model are spherical symmetry, no rotation, homogeneous density, and no pressures. A real star will have some small but non-vanishing quadrupole moment, it will have angular momentum, it will be composed of several kinds of gases with pressures and different equations of state, and its density will not be homogeneous. Therefore, it is reasonable to ask how gen-eral is the picture obtained in the OSD model and how much the collapse of a real star will depart from our mathematical idealization.

What happens to singularity and horizon once we introduce inhomogeneities, pres-sures, rotation, and asymmetries in the model? Gravitational collapse is a dynamical process, and the structure and evolution of the horizon in a realistic stellar interior is not well understood. We still do not have any analytical model of formation of a Kerr black hole. In fact, we do not even have any analytical solution describing a viable interior metric for the Kerr solution. Nevertheless, the evidence for the existence of black holes is now almost universally accepted, and most people believe that the process that leads to their formation can be roughly described via the OSD model. Still, the question of whether the black hole candidates that we observe in the universe are well described by the Schwarzschild and Kerr metrics is still open. As is the question of whether every collapsing star that is massive enough must inevitably form a black hole as the final end state. In order to study more realistic models, one needs to give up the hope to solve Einstein's equations analytically and resort to numerical simulations. Fully general relativistic simulations have been done in the past years to study (among other things) gravitational core collapse with rotation and magnetic fields, black hole mergers and black hole neutron star mergers, gravitational wave production, recoil from black hole mergers, production of jets, and gamma ray bursts (see, e.g., [24, 25, 42] and references therein).

Numerical simulations have improved dramatically over the last decade. Nevertheless, there are still no fully satisfactory simulations of supernovae explosions that lead to the formation of a black hole. One reason resides in the fact that many elements of classical and quantum physics come into play during the last stages of the life of a star. Describing accurately such scenarios is an enormous task that requires very expansive computations on the most advanced supercomputers. Further to this, numerical simulations must assume that GR holds unchanged at all energy scales and therefore are limited by our lack of knowledge of gravity in the strong field. Stellar evolution and black hole formation still present a lot of open questions, and the possibility exists that black holes are not the only possible final outcome of collapse of very massive stars. For these reasons, despite the increasing amount of observational evidence for the existence of black holes, it is useful to keep an eye open for other, more exotic, possibilities (see, e.g., [10, 16, 34, 45, 46] and references therein).

5.9   Concluding Remarks

Simple analytical models of general relativistic collapse show that a black hole can form as the end state of the life of a massive star. From a mathematical point of view, a black hole is a space-time singularity covered by an event horizon. The curvature singularity at the end of collapse is indicated by the divergence of the scalar $\mathcal{K}$, and approaching the singularity matter reaches infinite density in a finite co-moving time. In some sense, the singularity at the end of collapse is analogous to the infinite density obtained in Newtonian collapse and can be viewed as a limit of the model rather than a physical feature of the system. Singularities are found in many solutions of Einstein's equations. The question is whether they can form from physically realistic processes and how we should interpret their appearance. We may think that GR works well in the strong field regime and nothing can prevent complete collapse from happening. In this case, one has to accept that singularities are there and they may be causally connected to the outside universe. On the other hand, we can believe that GR works well in the strong field, but other effects arise either preventing the formation of singularities or hiding them from view. Or we can think that GR needs modifications in the strong field regime due perhaps to quantum effects. These modifications would then affect the space-time near the formation of classical singularities, thus removing them. The first attitude, although legitimate, is not very common. A singularity in any classical theory such as classical mechanics or electromagnetism is located somewhere in time and space and it does not affect the future predictability of space-time itself. On the other hand in GR, a singularity is not a part of the space-time. The distribution of matter determines the properties of space and time, and the occurrence of singularities translates in geodesic incompleteness and has important consequences for the causal structure of the space-time itself. For this reason, most people believe that singularities must not occur in the real universe. The second attitude can be summarized by the words of Roger Penrose [41]: *"...does there exist a 'cosmic censor' who forbids the appearance of naked singularities, clothing each one in an absolute event horizon?"* This is the famous cosmic censorship conjecture (CCC). At present, there exist counterexamples to the CCC, like the inhomogeneous dust collapse model, but their physical relevance is not entirely clear. On the other hand, it is highly plausible that GR is not enough to describe what happens in the strong field regime. One needs to account for microphysics or for modifications to GR possibly due to quantum effects. This third attitude is a view that was already suggested by Wheeler who saw singularities as possible probes for new physics.

If the occurrence of singularities at the end of collapse signals a breakdown of the fluid model approximation or a breakdown of GR itself, then what could be a better mathematical framework to describe the last stages of the life of a star? Is there any viable model for collapse that does not originate a singularity? The singularity theorems by Hawking and Penrose tell us that if GR is the ultimate ingredient that we need to use to describe collapse and if matter satisfies the usual energy conditions, then a singularity must necessarily form [19]. More precisely, provided that some energy condition is satisfied, the space-time is globally hyperbolic, and a trapped region develops at some point, a singularity must always form. Therefore, in order to develop non-singular model of collapse, one needs to modify GR in some way. Several attempts have been made over the years, and the general scenario that is arising is that singularities may be removed by quantum gravitational effects. Matter in the strong field regime may violate standard energy conditions, and the complete collapse to a black hole may be replaced by a bouncing scenario in which collapsing matter re-expands after reaching a minimal size. The expansion phase may take the form of an explosive event, and it may leave behind an exotic compact remnant (see, e.g., [3, 4, 17] and references therein).

These compact remnants may be less massive, more dense, and smaller than a neutron star and they would not possess an event horizon. Several types of exotic compact objects have been investigated, and their observational properties are of great interest for future astrophysical observations (see, e.g., [1, 2, 29, 30]). Given the small number of astrophysical black hole candidates observed so far and the peculiar features that theoretical compact objects may possess, it is reasonable to suppose that their observation may pose a great challenge for future astrophysics. Nevertheless, if some departure from the black hole paradigm will be observed in the future, this may open a window onto new areas of physics where gravitation and quantum mechanics merge.

References

1. C. Bambi, Phys. Rev. D **87**, 023007 (2013)
2. C. Bambi, D. Malafarina, Phys. Rev. D **88**, 064022 (2013)
3. C. Bambi, D. Malafarina, L. Modesto, Phys. Rev. D **88**, 044009 (2013)
4. M. Bojowald, R. Goswami, R. Maartens, P. Singh, Phys. Rev. Lett. **95**, 091302 (2005)
5. K. Bolejko, A. Krasiński, C. Hellaby, M.N. Celerier, *Structures in the Universe by Exact Methods: Formation, Evolution, Interactions*, sect. 18 (Cambridge University Press, Cambridge, 2010)
6. H. Bondi, Mon. Not. Astron. Soc. **107**, 343 (1947)
7. I. Booth, Can. J. Phys. **83**, 1073 (2005)
8. S. Chandrasekhar, Asrophys. J. **74**, 81 (1931)
9. D. Christodoulou, Commun. Math. Phys. **93**, 171 (1984)
10. M. Colpi, S.L. Shapiro, I. Wasserman, Phys. Rev. Lett. **57**, 2485–2488 (1986)
11. S. Datt, Zs. F. Phys. **108**, 314 (1938)
12. D.M. Eardley, L. Smarr, Phys. Rev. D **19**, 2239 (1979)
13. F. Fayos, X. Jaen, E. Llanta, J.M.M. Senovilla, Phys. Rev. D **45**, 2732 (1992)
14. F. Fayos, J.M.M. Senovilla, R. Torres, Phys. Rev. D **54**, 4862 (1996)
15. D. Finkelstein, Phys. Rev. **110**, 965 (1958)
16. B. Freedman, L.D. McLerran, Phys. Rev. D **17**, 11091122 (1978)
17. R. Goswami, P.S. Joshi, P. Singh, Phys. Rev. Lett. **96**, 031302 (2006)
18. S.W. Hawking, G.F.R. Ellis, *The Large Scale Structure of Space-time* (Cambridge University Press, Cambridge, 1973)
19. S.W. Hawking, R. Penrose, Proc. R. Soc. Lond. A **314**, 529 (1970)
20. S.A. Hayward, Phys. Rev. D **49**, 6467 (1994)
21. C. Hellaby, K. Lake, Astrophys. J. **290**, 381 (1985)
22. C. Hellaby, K. Lake, Astrophys. J. **300**, 461 (1986)
23. W. Israel, Nuovo Cimento B **44**, 1 (1966); Nuovo Cimento B **48**, 463 (1966)
24. H.T. Janka, Annu. Rev. Nucl. Part. Sci. **62**(1), 407 (2012)
25. H.T. Janka, F. Hanke, L. Hdepohl, A. Marek, B. Mller, M. Obergaulinger, Prog. Theor. Exp. Phys. **2012**(1), id.01A309 (2012)
26. P.S. Joshi, I.H. Dwivedi, Phys. Rev. D **47**, 5357 (1993)
27. P.S. Joshi, I.H. Dwivedi, Class. Quantum Gravity **16**, 41 (1999)
28. P.S. Joshi, D. Malafarina, Int. J. Mod. Phys. D **20**(14), 2641 (2011)
29. P.S. Joshi, D. Malafarina, R. Narayan, Class. Quantum Gravity **28**, 235018 (2011)
30. P.S. Joshi, D. Malafarina, R. Narayan, Class. Quantum Gravity **31**, 015002 (2014)
31. A. Krasinski, *Inhomogeneous Cosmological Models* (Cambridge University Press, Cambridge, 1997)
32. A. Krasinski, J. Plebanski, *Introduction to General Relativity and Cosmology*, sect. 18.15 (Cambridge University Press, Cambridge, 2006), pp. 301
33. G. Lemaître, Ann. Soc. Sci. Bruxelles I, A **53**, 51 (1933)
34. P.O. Mazur, E. Mottola (2001), arXiv:0109035 [gr-qc]
35. C. Misner, D. Sharp, Phys. Rev. **136**, B571 (1964)
36. C.W. Misner, K.S. Thorne, J.A. Wheeler, *Gravitation* (W. H Freeman, San Francisco, 1973)

37. R.P.A.C. Newman, Class. Quantum Gravity **3**, 527 (1986)
38. J.R. Oppenheimer, H. Snyder, Phys. Rev. **56**, 455 (1939)
39. J.R. Oppenheimer, G.M. Volkov, Phys. Rev. **56**, 374 (1939)
40. R. Penrose, Phys. Rev. Lett. **14**, 57 (1965)
41. R. Penrose, Rivista del Nuovo Cimento **1**, 257 (1969)
42. L. Rezzolla, B. Giacomazzo, L. Baiotti, J. Granot, C. Kouveliotou, M.A. Aloy, Astrophys. J. Lett. **732**, L6 (2011)
43. R.C. Tolman, Proc. Natl. Acad. Sci. USA **20**, 410 (1934)
44. R.C. Tolman, Phys. Rev. **55**, 364 (1939)
45. D.F. Torres, S. Capozziello, G. Lambiase, Phys. Rev. D **62**, 104012 (2000)
46. M. Visser, C. Barcelo, S. Liberati, S. Sonego (2009), arXiv:0902.0346 [gr-qc]
47. R.M. Wald, *General Relavitity*, sect. 6.2 (University of Chicago Press, Chicago, 1984), p. 125
48. B. Waugh, K. Lake, Phys. Rev. D **38**, 1315 (1988)
49. P. Yodzis, H.-J. Seifert, H. Muller zum Hagen, Commun. Math. Phys. **34**, 135 (1973)
50. J.B. Zeldovich, L.F. Grishchuk, Mon. Not. R. Astron. Soc. **2**(07), 23 (1984)

Appendix A
General Relativity in a Nutshell

The aim of this appendix was to provide the reader not familiar with general relativity a basic theoretical background on some fundamental concepts. It is not an introduction to the theory of general relativity, but is hopefully enough to understand the chapters in this volume. A key concept is the existence of an innermost stable circular orbit around a black hole, which has no Newtonian counterpart. Since we are interested in astrophysics, here we always assume that the spacetime has 4 dimensions $(1 + 3)$. Units in which $G_N = c = 1$ are used, and therefore, all the length and timescales are set by the black hole mass M. The conversion factors are

$$M = 1.477 \left(\frac{M}{M_\odot} \right) \text{ km}, \quad M = 4.925 \left(\frac{M}{M_\odot} \right) \mu\text{s}. \tag{A.1}$$

A.1 Geodesic Equations

In classical mechanics, the principle of least action plays a very important role and it can be used to obtain in an elegant way the equations of motion for a system when its action is known. In the case of a free point-like particle, the Lagrangian is simply given by the kinetic energy of the particle

$$L = \frac{1}{2}mv^2 = \frac{1}{2}mg_{ij}\frac{dx^i}{dt}\frac{dx^j}{dt}, \tag{A.2}$$

where m is the mass of the particle, g_{ij} is the metric tensor, $\{x^i\}$ are the coordinates, and t is the time. Here, we use the Einstein convention of summation over repeated indices; that is,

$$g_{ij}\frac{dx^i}{dt}\frac{dx^j}{dt} = \sum_{i,j} g_{ij}\frac{dx^i}{dt}\frac{dx^j}{dt}. \tag{A.3}$$

© Springer-Verlag Berlin Heidelberg 2016
C. Bambi (ed.), *Astrophysics of Black Holes*, Astrophysics
and Space Science Library 440, DOI 10.1007/978-3-662-52859-4

For instance, in Cartesian coordinates $\{x, y, z\}$, we have

$$v^2 = \left(\frac{dx}{dt}\right)^2 + \left(\frac{dy}{dt}\right)^2 + \left(\frac{dz}{dt}\right)^2, \tag{A.4}$$

which means that the only nonvanishing coefficients of the metric tensor are $g_{xx} = g_{yy} = g_{zz} = 1$, while all the off-diagonal ones are zero. In the case of spherical coordinates $\{r, \theta, \phi\}$, we have

$$v^2 = \left(\frac{dr}{dt}\right)^2 + r^2 \left(\frac{d\theta}{dt}\right)^2 + r^2 \sin^2 \theta \left(\frac{d\phi}{dt}\right)^2, \tag{A.5}$$

so $g_{rr} = 1$, $g_{\theta\theta} = r^2$, $g_{\phi\phi} = r^2 \sin^2 \theta$, and all the other coefficients vanish. It is easy to see that if we want to go from a coordinate system $\{x^i\}$ to another one $\{x'^i\}$, the metric tensor changes as

$$g_{ij} \rightarrow g'_{ij} = \frac{\partial x^a}{\partial x'^i} \frac{\partial x^b}{\partial x'^j} g_{ab}. \tag{A.6}$$

Infinitesimal displacements $\{dx^i\}$ change in the opposite way; that is,

$$dx^i \rightarrow dx'^i = \frac{\partial x'^i}{\partial x^a} dx^a, \tag{A.7}$$

and for this reason we write lower indices for g_{ij} and upper index for dx^i.

For what follows, it is more convenient to define the action of a free point-like particle as proportional to the length of its path

$$S = m \int_\gamma ds = m \int_\gamma L \, d\lambda, \tag{A.8}$$

where m is the particle mass, ds is the line element,

$$L = \sqrt{g_{ij} \dot{x}^i \dot{x}^j}, \tag{A.9}$$

λ is an affine parameter that parameterizes the particle path $\gamma(\lambda)$ and the dot indicates the derivative with respect to λ. From the principle of least action, we find the Euler–Lagrange equations

$$\frac{d}{d\lambda} \frac{\partial L}{\partial \dot{x}^i} - \frac{\partial L}{\partial x^i} = 0. \tag{A.10}$$

It is easy to see that Eqs. (A.2) and (A.9) provide the same equations of motion. If we plug Eq. (A.9) into (A.10), we obtain the geodesic equations

$$\ddot{x}^i + \Gamma^i_{jk}\dot{x}^j\dot{x}^k = 0, \tag{A.11}$$

where Γ^i_{jk} are the Christoffel symbols

$$\Gamma^i_{jk} = \frac{1}{2}g^{il}\left(\frac{\partial g_{lk}}{\partial x^j} + \frac{\partial g_{jl}}{\partial x^k} - \frac{\partial g_{jk}}{\partial x^l}\right). \tag{A.12}$$

In Cartesian coordinates, all the Christoffel symbols vanish, and therefore, the geodesic equations simply reduce to $\ddot{x} = \ddot{y} = \ddot{z} = 0$.

In special relativity, time and space are not independent entities and the line element of the spacetime ds is given, respectively, in Cartesian and spherical coordinates, by

$$ds^2 = -dt^2 + dx^2 + dy^2 + dz^2, \tag{A.13}$$
$$ds^2 = -dt^2 + dr^2 + r^2 d\theta^2 + r^2 \sin^2\theta d\phi^2. \tag{A.14}$$

Thanks to the transformation rules (A.6) and (A.7) the line element is an invariant; that is, it is independent of the choice of the coordinates. We can thus define the following coordinate independent types of trajectories:

$$ds^2 < 0 \quad \text{timelike trajectories,}$$
$$ds^2 = 0 \quad \text{lightlike trajectories,}$$
$$ds^2 > 0 \quad \text{spacelike trajectories.} \tag{A.15}$$

In particular, massless particles such as photons will follow light-like trajectories with $ds^2 = 0$. In the case of massive particles, it is convenient to use their "proper time" τ, i.e., the time measured in the rest-frame of the particle, as the affine parameter λ. Since ds^2 is an invariant, $d\tau^2 = -ds^2$, because the coordinate system is anchored on the particle and therefore, there is no motion along the spatial directions. With this choice of the affine parameter, $ds^2 = -1$. Since we are now considering a spacetime in $1+3$ dimensions, it is common to use Greek letters $\mu, \nu, \rho, \ldots$ to denote spacetime indices, for instance, $g_{\mu\nu}$, with $\mu = 0, 1, 2$, and 3, where 0 stands for the temporal component, while 1, 2, and 3 for the spacial components. Latin letter $i, j, k, \ldots$ are used for the space components and therefore can assume the values 1, 2, and 3.

In Newtonian mechanics, the motion of a test particle in a gravitational field can be described by adding the correct gravitational potential to the Lagrangian of the free particle. A key point in general relativity is that the gravitational field can be absorbed into the metric tensor $g_{\mu\nu}$: In other words, we have still a free particle, but now it lives in a curved spacetime and follows the geodesics of that spacetime. It is useful to see how we can recover the Newtonian limit. In Cartesian coordinates, the metric tensor of the flat spacetime of special relativity is usually denoted by $\eta_{\mu\nu}$, where [see Eq. (A.13)]

$$||\eta_{\mu\nu}|| = \text{diag}(-1, 1, 1, 1). \tag{A.16}$$

The Newtonian limit should be recovered by requiring that: (i) The gravitational field is weak, (ii) the gravitational field is stationary, and (iii) the motion of the particle is nonrelativistic. These three conditions are given, respectively, by

$$g_{\mu\nu} = \eta_{\mu\nu} + h_{\mu\nu} \quad \text{with } |h_{\mu\nu}| \ll 1, \tag{A.17}$$

$$\frac{\partial g_{\mu\nu}}{\partial t} = 0, \tag{A.18}$$

$$\frac{dt}{d\lambda} \gg \frac{dx^i}{d\lambda}. \tag{A.19}$$

Within these approximations, the geodesic equations reduce to

$$\frac{d^2 x^\mu}{d\lambda^2} + \Gamma^\mu_{tt} \left(\frac{dt}{d\lambda}\right)^2 = 0 \quad \text{with } \Gamma^\mu_{tt} = \frac{1}{2}\eta^{\mu\nu}\frac{\partial h_{tt}}{\partial x^\nu}. \tag{A.20}$$

After a simple integration, we find

$$\frac{d^2 x^i}{dt^2} = -\frac{1}{2}\frac{\partial h_{tt}}{\partial x^i}. \tag{A.21}$$

If we compare Eq. (A.21) with the Newtonian formula $m\ddot{\mathbf{x}} = -m\nabla\Phi$, where Φ is the Newtonian gravitational potential, and we require that the spacetime is flat at infinity, we find

$$g_{tt} = -(1 + 2\Phi). \tag{A.22}$$

A.2 Einstein Equations

In the previous section, we have discussed the motion of test particles in a given background metric $g_{\mu\nu}$. In general relativity, the latter takes into account the gravitational field as well, and therefore, it is determined by the matter distribution. The Einstein equations relate the spacetime geometry (on the left side) to the matter content (on the right side):

$$G_{\mu\nu} = 8\pi T_{\mu\nu}, \tag{A.23}$$

where $G_{\mu\nu}$ is the Einstein tensor

$$G_{\mu\nu} = R_{\mu\nu} + \frac{1}{2}g_{\mu\nu}R, \tag{A.24}$$

$R_{\mu\nu}$ and R are, respectively, the Ricci tensor and the scalar curvature

$$R_{\mu\nu} = \frac{\partial \Gamma^{\rho}_{\mu\nu}}{\partial x^{\rho}} - \frac{\partial \Gamma^{\rho}_{\mu\nu}}{\partial x^{\nu}} + \Gamma^{\sigma}_{\mu\nu}\Gamma^{\rho}_{\sigma\rho} - \Gamma^{\sigma}_{\mu\rho}\Gamma^{\rho}_{\nu\sigma}, \tag{A.25}$$

$$R = g^{\mu\nu}R_{\mu\nu}, \tag{A.26}$$

and $T_{\mu\nu}$ is the matter energy-momentum tensor. The factor 8π is just to recover the correct Newtonian limit. If the matter content is known, we can plug its energy-momentum tensor on the right-hand side of the Einstein equations and find the corresponding metric tensor (modulo a choice of coordinates). However, this job is far from being trivial, mainly because of the nonlinear nature of the Einstein equations. Analytical solutions are thus known only in the case of special symmetries.

To find the Newtonian limit, we assume the approximations (A.17) and (A.18), as well as that in our coordinate system all the components of the matter energy-momentum tensor are negligible, except the tt one, which describes the energy density and reduces to the matter density in the Newtonian limit, so that

$$T_{tt} = \rho \quad T_{\mu\nu} = 0 \ \text{for} \ \mu \neq \nu \neq t. \tag{A.27}$$

After some calculations, we find

$$R_{tt} = \frac{1}{2}\Delta h_{tt} = 8\pi\rho, \tag{A.28}$$

where Δ is the Laplace operator. The Poisson equation of Newtonian gravity is recovered by replacing h_{tt} with 2Φ, where Φ is the Newtonian gravitational potential, as found in the previous section.

A.3 Schwarzschild Solution

The only spherically symmetric solution of the vacuum Einstein equations $G_{\mu\nu} = 0$ is the Schwarzschild metric (Birkhoff's theorem). Such a solution describes the exterior gravitational field of any spherically symmetric source, which therefore may also have a nonvanishing radial motion. The line element is

$$ds^2 = -\left(1 - \frac{2M}{r}\right)dt^2 + \left(1 - \frac{2M}{r}\right)^{-1}dr^2 + r^2 d\theta^2 + r^2 \sin^2\theta d\phi^2, \tag{A.29}$$

where M is a free parameter to be related to the gravitational mass of the object. For $M/r \ll 1$, the correct Newtonian limit is recovered. In the case of a star, this metric is valid in the exterior, for radii $r > R_{\star}$, where $R_{\star}$ is the radial coordinate of the surface of the star. In the case of a black hole, there is no $R_{\star}$. $r = 2M$ is the radius of the event horizon, where the above coordinate system is ill defined. Since $g_{rr} \to +\infty$ for $r \to 2M$, from the geodesic equations we find that any massless and massive particle in the region $r \leq 2M$ is trapped and cannot escape to infinity. The event horizon is the boundary of the black hole and it can be seen as a one-way membrane: particles can cross the event horizon from the outside, and thus be swallowed by the black hole, but once inside they cannot come back to the outside region.

The event horizon is a region with quite peculiar properties. For instance, the relation between the temporal coordinate t and the proper time of a static observer at (r, θ, ϕ) is

$$d\tau = \left(1 - \frac{2M}{r}\right)^{1/2} dt < dt. \tag{A.30}$$

The observer proper time τ is thus slower than the coordinate time t (corresponding to the proper time of a static observer at infinity) as a consequence of the gravitational field, and the observer proper time is frozen out (with respect to the distant observer) for $r = 2M$.

Let us now consider two static observers, say A and B, with coordinates, respectively, (r_A, θ, ϕ) and (r_B, θ, ϕ). The observer A emits an electromagnetic signal that he/she measures to have frequency ν_A and to last for a time $\Delta\tau_A$. The number of wavefronts is thus $n = \nu_A \Delta\tau_A$. The signal arrives at the position of the observer B, who measures a frequency ν_B for a time $\Delta\tau_B$. As the number of wavefronts is an invariant, we have $\nu_A/\nu_B = \Delta\tau_B/\Delta\tau_A$. For a light ray, $ds^2 = 0$, and therefore, since A and B have the same θ and ϕ coordinates, along the photon path we have

$$\left(1 - \frac{2M}{r}\right) dt^2 = \left(1 - \frac{2M}{r}\right)^{-1} dr^2. \tag{A.31}$$

The time interval with respect to the coordinate system that the first wavefront takes to go from the observer A to the observer B is

$$t_B^1 - t_A^1 = \int_{r_A}^{r_B} \frac{dr}{1 - 2M/r}. \tag{A.32}$$

Since it is independent of the coordinate t, the last wavefront takes the same time; that is, $t_B^1 - t_A^1 = t_B^n - t_A^n$ and therefore, $t_A^n - t_A^1 = t_B^n - t_B^1$. The proper time measured by the observer A is instead

$$\Delta\tau_A = \int_{t_A^1}^{t_A^n} \left(1 - \frac{2M}{r_A}\right)^{1/2} dt = \left(1 - \frac{2M}{r_A}\right)^{1/2} \left(t_A^n - t_A^1\right), \tag{A.33}$$

and the same for B, with the index B replacing the index A in Eq. (A.33). We can then find the relation between ν_A and ν_B

$$\frac{\nu_A}{\nu_B} = \left(1 - \frac{2M}{r_B}\right)^{1/2} \left(1 - \frac{2M}{r_A}\right)^{-1/2}. \tag{A.34}$$

Even in this case, we can see that these coordinates are not suitable to describe what happens at $r = 2M$.

In astrophysical scenarios, one usually needs to compute the trajectories of particles and photons in this background. In the case of an accretion disk, we may imagine that the particles of the gas follow nearly geodesic circular orbit in the equatorial plane $\theta = \pi/2$. Since the metric coefficients in Eq. (A.29) are independent of the coordinates t and ϕ, there are two constants of motions, associated, respectively, with the energy and the angular momentum of the particle

$$\frac{d}{d\lambda}\frac{\partial L}{\partial \dot{t}} = 0 \quad \Rightarrow \quad \left(1 - \frac{2M}{r}\right)\dot{t} = E = \text{const.,} \qquad (A.35)$$

$$\frac{d}{d\lambda}\frac{\partial L}{\partial \dot{\phi}} = 0 \quad \Rightarrow \quad r^2\dot{\phi} = L = \text{const.,} \qquad (A.36)$$

For a massive test particle $ds^2 = g_{\mu\nu}\dot{x}^\mu\dot{x}^\nu = -1$ and we have

$$\frac{1}{2}\dot{r}^2 = \frac{E^2 - 1}{2} + \frac{M}{r} - \frac{L^2}{r^2} + \frac{2ML^2}{r^3}. \qquad (A.37)$$

Equation (A.37) can be seen as the equation of motion of a test particle in Newtonian mechanics under the effect of an effective potential. The first term on the right-hand side is just a constant. The second term, M/r, is the standard gravitational potential of a spherically symmetric mass in the Newtonian theory and it introduces an attractive force (the term is positive). The third term, L^2/r^2, is the standard centrifugal potential and introduces an effective repulsive force (the term is negative). The forth term, $2ML^2/r^3$, is something absent in the Newtonian theory: It becomes dominant at very small radii (it scales as $1/r^3$) and it is attractive. Such a term introduces some novel properties, like the fact that circular orbits exist only for radii larger than a critical one, say $r \geq r_c$, and that stable circular orbits exist only for radii larger than another critical value, $r \geq r_{\text{ISCO}}$. These two facts can be understood by noticing that the term $2ML^2/r^3$ makes the gravitational force very strong at small radii, and therefore, any particle must fall onto the black hole. More details will be given in the next section.

A.4   Kerr Solution

The only rotating uncharged black hole solution of 4-dimensional general relativity is described by the Kerr metric, which depends only on the black hole mass M and the black hole spin angular momentum J. This is the statement of the "no-hair" theorem, which is called in this way to mean that a black hole has no distinguishing features (no hair), except for its mass and spin angular momentum. An astrophysical black hole is supposed to be well described by the Kerr solution. In general relativity, initial deviations from the Kerr metric are quickly radiated away through the emission of gravitational waves. The equilibrium electric charge is soon reached, because of the highly ionized host environment, and too small for macroscopic objects to affect the geometry of the spacetime. The effect of the mass of the accretion disk is completely negligible in most cases.

In Boyer–Lindquist coordinates, the line element is

$$ds^2 = -\left(1 - \frac{2Mr}{\Sigma}\right)dt^2 - \frac{4aMr\sin^2\theta}{\Sigma}dt\,d\phi$$
$$+\frac{\Sigma}{\Delta}dr^2 + \Sigma d\theta^2 + \left(r^2 + a^2 + \frac{2a^2Mr\sin^2\theta}{\Sigma}\right)\sin^2\theta\,d\phi^2, \quad (A.38)$$

where $a = J/M$, $\Sigma = r^2 + a^2\cos^2\theta$ and $\Delta = r^2 - 2Mr + a^2$. The radius of the event horizon is defined by the larger root of $\Delta = 0$, where g_{rr} diverges, and it is

$$r_H = M + \sqrt{M^2 - a^2}. \quad (A.39)$$

Let us note that the event horizon exists only for $|a| \leq M$. For $|a| > M$, there is no horizon and the Kerr metric describes the spacetime of a naked singularity. A number of arguments (instability of the spacetime, apparent impossibility to create a Kerr naked singularity from a Kerr black hole, etc.) suggest that the Kerr metric with $|a| > M$ has no astrophysical applications. An important new property of the Kerr metric with respect to the Schwarzschild solution is the existence of the ergosphere, where g_{tt} changes sign and static particles are not allowed (everything must rotate). The outer boundary of the ergosphere is

$$r_E = M + \sqrt{M^2 - a^2\cos^2\theta}. \quad (A.40)$$

As in the Schwarzschild case, the metric coefficients are independent of the coordinates t and ϕ, and therefore, we have two constants of motion, associated, respectively, with the energy E and the axial component of the angular momentum L_z. We can exploit this fact to write $\dot{t}$ and $\dot{\phi}$ as

$$\dot{t} = \frac{Eg_{\phi\phi} + L_z g_{t\phi}}{g_{t\phi}^2 - g_{tt}g_{\phi\phi}}, \qquad \dot{\phi} = -\frac{Eg_{t\phi} + L_z g_{tt}}{g_{t\phi}^2 - g_{tt}g_{\phi\phi}}. \quad (A.41)$$

From $g_{\mu\nu}\dot{x}^\mu\dot{x}^\nu = -1$, we can write

$$g_{rr}\dot{r}^2 + g_{\theta\theta}\dot{\theta}^2 = V_{\text{eff}}(r, \theta), \quad (A.42)$$

where the effective potential V_{eff} is given by

$$V_{\text{eff}} = \frac{E^2 g_{\phi\phi} + 2EL_z g_{t\phi} + L_z^2 g_{tt}}{g_{t\phi}^2 - g_{tt}g_{\phi\phi}} - 1. \quad (A.43)$$

Circular motion on the equatorial plane plays an important role in the description of accretion disks because, even if there is an initial misalignment, the disks are forced to adjust on the equatorial plane (Bardeen–Petterson effect). Circular orbits on the equatorial plane are located at the zeros and the turning points of the effective potential: $\dot{r} = \dot{\theta} = 0$, which implies $V_{\rm eff} = 0$, and $\ddot{r} = \ddot{\theta} = 0$, requiring, respectively, $\partial_r V_{\rm eff} = 0$ and $\partial_\theta V_{\rm eff} = 0$. From these conditions, one can obtain the values of E and L_z

$$E = \frac{r^{3/2} - 2Mr^{1/2} \pm aM^{1/2}}{r^{3/4}\sqrt{r^{3/2} - 3Mr^{1/2} \pm 2aM^{1/2}}}, \tag{A.44}$$

$$L_z = \pm \frac{M^{1/2}\left(r^2 \mp 2aM^{1/2}r^{1/2} + a^2\right)}{r^{3/4}\sqrt{r^{3/2} - 3Mr^{1/2} \pm 2aM^{1/2}}}. \tag{A.45}$$

The orbits are stable under small perturbations if $\partial_r^2 V_{\rm eff} \leq 0$ and $\partial_\theta^2 V_{\rm eff} \leq 0$. In Kerr spacetime, the second condition is always satisfied, so one can deduce the radius of the innermost stable circular orbit (ISCO) from $\partial_r^2 V_{\rm eff} = 0$. After some passages, one finds the ISCO radius

$$r_{\rm ISCO} = 3M + Z_2 \mp \sqrt{(3M - Z_1)(3M + Z_1 + 2Z_2)}, \tag{A.46}$$

$$Z_1 = M + \left(M^2 - a^2\right)^{1/3}\left[(M + a)^{1/3} + (M - a)^{1/3}\right],$$

$$Z_2 = \sqrt{3a^2 + Z_1^2}, \tag{A.47}$$

The ISCO radius is $r_{\rm ISCO} = 6M$ for a non-rotating black hole and decreases (increases) as the spin parameter increases for a corotating (counter-rotating) disk, to $r_{\rm ISCO} = M$ ($r_{\rm ISCO} = 9M$) for a maximally rotating Kerr black hole with $a = M$.

Printed in the United States
By Bookmasters